Gregor Lax

From Atmospheric Chemistry to Earth System Science

GREGOR LAX

FROM ATMOSPHERIC CHEMISTRY TO EARTH SYSTEM SCIENCE

CONTRIBUTIONS TO THE RECENT HISTORY OF THE MAX PLANCK INSTITUTE FOR CHEMISTRY (OTTO HAHN INSTITUTE), 1959 – 2000

DIEPHOLZ · BERLIN 2018
GNT-Verlag

BIBLIOGRAPHIC INFORMATION PUBLISHED BY THE
DEUTSCHE NATIONALBIBLIOTHEK
The Deutsche Nationalbibliothek lists this publication in the Deutsche Nationalbibliografie; detailed bibliographic data are available on the Internet at http://dnb.dnb.de.

Translation and edited version of the preprint "Von der Atmosphärenchemie zur Erforschung des Erdsystems. Beiträge zur jüngeren Geschichte des Max-Planck-Instituts für Chemie (Otto-Hahn-Institut), 1959 – 2000." Published 2018 in "Forschungsprogramm Geschichte der Max-Planck-Gesellschaft (gmpg)" at the Max Planck Institute for the History of Science.
gmpg.mpiwg-berlin.mpg.de

TRANSLATION
Dr. Sabine Wacker, Aichwald, Germany
www.wacker-translation.de

COPY EDITING
Dr. Tracey Andreae

EDITOR
Max Planck Institute for Chemistry (Otto Hahn Institute),
Hahn-Meitner-Weg 1, 55128 Mainz, Germany
www.mpic.de

PUBLISHER
GNT-Verlag GmbH, Schloßstr. 1, 49356 Diepholz, Germany
www.gnt-verlag.de

ISBN 978-3-86225-112-4
Printed by KDD Kompetenzzentrum Digital-Druck GmbH, Nuremberg, Germany.

TABLE OF CONTENTS

PREFACE

The phrase *"Earth System"* first became important in research and research-policy contexts in 1983 because of its use by the NASA *Earth System Committee*. This represented a major step towards initiating "global change" research, which has had a long-term influence on national and international research programs and organizations. The "Earth System" concept, nowadays a well-known buzzword, gained extensive popularity in the context of the large *global change* programs, for example the *World Climate Research Program* (WCRP) and the *International Geosphere Biosphere Program* (IGBP).[1] The *Earth System Sciences* (ESS) are comparatively new and examine the interactions and mutual influences in and between the Earth's subsystems (biosphere, geosphere, cryosphere, hydrosphere, atmosphere etc.). The formation and consolidation of ESS as an independent discipline can be regarded as having been largely completed by the mid-2000s in Germany. This is based on the assumption that the following parameters can be used for the formation and consolidation of an independent academic discipline: 1. Large-scale funding in connection with research programs and the foundation of new institutions. 2. The establishment of professorships and courses of study, along with the associated education of young academic talent. 3. The formation of specific journals and a canonization process for the specialist literature.[2]

1 Cf. Uhrqvist/Lövbrand, Rendering global change.

2 Schützenmeister described this, taking atmospheric research as an example. Cf. Schützenmeister, Zwischen Problemorientierung, 109 f.

In Germany today, all major institutions and organizations of the German science system are involved, e. g. the Leopoldina,[3] the Max Planck Society (MPG),[4] the Helmholtz Association (HGF),[5] Leibniz Institutes, such as the Institute for Tropospheric Research in Leipzig (TROPOS),[6] the German Research Foundation (DFG)[7] and several universities, such as Mainz, Hamburg, Bremen, and Hohenheim.

The origins of the ESS however, do not lie in the Earth System concept which began in the early 1980s, but rather in the atmospheric sciences rapidly emerging in the mid 1950s in particular in the US and Sweden, whose approaches developed from the primary observation of atmospheric phenomena to the examination of interactive relationships and exchange processes between the atmosphere and other spheres of the Earth (biosphere, geosphere, cryosphere etc.). This integrative perspective resulted in particular in the long-term formation of climate research at both an organizational and epistemic level, and which since the second half of the 1950s has been mainly influenced by opinions that were mechanistic and decidedly based on atmospheric *chemistry*. These opinions gradually began to gain hold in the FRG only in the late 1960s, with a considerable latency compared with the US, for example, and significantly enhanced classic German meteorology, which previously was primarily geared towards weather phenomena and weather forecasts and lagged behind international developments by a good decade.

---

3 See the Internet presence of the Leopoldina's Earth System Research Working Group: <https://www.leopoldina.org/politikberatung/arbeitsgruppen/erdsystemforschung/>. Status: May 23, 2018.

4 Andreae/Marotzke/Heimann, Partnerschaft Erdsystemforschung.

5 Cf. Helmholtz-Gemeinschaft, Helmholtz-Roadmap.

6 See TROPOS' description of the research of dust sources as given on their Internet presence: <http://www.tropos.de/institut/abteilungen/modellierung-atmosphaeri›scher-prozesse/transportprozesse/staubquellen/>. Status: May 23, 2018.

7 See the description given by the Senate Commission on the DFG website: <http://www.dfg.de/dfg_profil/gremien/senat/erdsystemforschung/>. Status: March 9, 2018.

Both the establishment of new integrative approaches in German atmospheric research and the history of atmospheric and Earth System sciences as a whole are inextricably linked to the Max Planck Society. Starting in 1968, and for roughly the next four decades, an Earth System cluster with a steadily growing personnel and institutional network was formed in the MPG. At the epistemic level, it forced the use of specific approaches and methods. At the science-policy level, it gained significant influence both within the MPG and in the FRG and the international scientific community.

Central pillars of this process were the establishment of a department for atmospheric chemistry at the MPI for Chemistry in Mainz (MPIC) in 1968, under the leadership of meteorologist Christian Junge, and the founding of the MPI for Meteorology (MPI-M) in Hamburg 1975, and finally the MPI for Biogeochemistry (MPI-BGC) in Jena in 1996/1997. Alongside these three major institutes, there were other facilities that took, or still take, Earth System approaches at a department level. An example of this is the department for cosmophysics at the MPI for Nuclear Physics in Heidelberg. Between 1994 and 2003 there were two directors there. One of them was Konrad Mauersberger, who led the group for atmospheric physics that took approaches that were clearly Earth-system-based. During his term, Mauersberger was a member of almost all commissions that dealt with appointments and topic areas at or relating to the MPIs for Chemistry, Meteorology and Biogeochemistry. One particular visible manifestation of the Earth System cluster at the MPG came in the form of the "Earth System Research" partnership that was established in 2006.[8] This initiative currently represents the MPG's Earth System research cluster both internally and externally, and functions as a coordinating forum, information portal, and shared presence.

The subject matter covered by the present observations is the settlement, establishment and expansion of atmospheric and Earth System

8 Andreae/Marotzke/Heimann, Partnerschaft Erdsystemforschung.

science research at the Max Planck Institute for Chemistry in Mainz. Within the MPG, the institute is both the origin and one of the pillars for these areas. The overall history of the formation of the Earth System cluster at the MPG, which has spanned roughly four decades, is part of the program for the history of the Max Planck Society (GMPG), which was initiated in 2014 and is based at the Max Planck Institute for the History of Science. In this program, additional work will be carried out in connection with an "Earth System Sciences" subproject which commenced in January of this year. The present observations are to be considered in this regard also as foundations in the context of this subproject, contributing to the overall historical understanding of the development of ESS at the MPG, in the scientific landscape in the Federal Republic of Germany and in the national and international scientific community.

INTRODUCTION

The Max Planck Institute for Chemistry in Mainz (MPIC) has an eventful history that now spans more than one hundred years. In 1912, the institute opened as the Kaiser Wilhelm Institute for Chemistry (KWIC) in Berlin-Dahlem and together with the MPI for Physical Chemistry and Electrochemistry (Fritz Haber Institute) it is one of the two oldest institutions of the Kaiser Wilhelm Society that was re-founded in 1948 as the Max Planck Society. Due to heavy damage from bomb hits in 1944, the institute was evacuated to Tailfingen in the Swabian Alb (Albstadt, Baden-Württemberg)[9] and subsequently moved to Mainz in Rhineland-Palatinate, primarily under the leadership of the second director and one of the co-discoverers of nuclear fission, Fritz Straßmann (1902 – 1980). The MPIC remained at that location, on the campus of the Johannes Gutenberg University in Mainz that had reopened in 1946 on the site of former French military barracks, until 2012. The most recent move then brought the institute to its current new building, a few streets further away, located at Hahn-Meitner-Weg 1.

Over its history, researchers at the institute have worked in fields ranging from organic chemistry in the 1910s, to radiochemistry (until the mid-1950s) and physical chemistry (until the end of the 1970s), up to cosmo-, geo-, atmospheric, and biogeochemistry. In the course of the 20th century, the MPIC has been connected with the names of numerous prominent scientists, including Nobel Laureates Emil

9 For details on the episode in Tailfingen, see the monograph: Lässing, Teufel.

Fischer (1852 – 1919), Richard Willstätter (1872 – 1942), Otto Hahn (1879 – 1968), and Paul Crutzen (born 1933). The range of research topics over the century spans color chemistry and the discovery of nuclear fission, examinations of moon rocks brought back by Apollo 11, research on the chemical composition of the Earth's atmosphere, and finally, investigations of mutual chemical processes in the Earth System as a whole.

As one of the oldest and most renowned non-university research institutions in Germany, the MPIC represents a rich resource for historical scientific research. Until now, however, the historiography has primarily addressed the early history[10]; more recent episodes in the history of the institute and questions about its role in the context of the scientific landscape in the Federal Republic have largely remained unexplored—with the exception of a commemorative publication on the occasion of the one hundredth anniversary, published in 2012 with some contributions that also addressed aspects of more recent times.[11]

Although certainly desirable, a complete presentation of the history of the institute since the foundation of the FRG should not and cannot be presented here. Instead, this study addresses specific aspects of the institute's history, with particular focus on atmospheric chemistry research that was established there at the end of the 1960s and continues today. The current research orientation of the MPIC on broad chemical processes of the Earth System can scarcely be understood without this background. Core fields of research on atmospheric and Earth System have been or are being examined and at times re-explored. The spectrum of research ranges from the exploration of atmospheric trace gases, anthropogenic influences, greenhouse gases and biomass combustion, to theories of nuclear winter, the study and explanation of the ozone hole, and the suggestion of the name "Anthropocene"

10 To name just a few examples: Johnson, Kaiser's Chemists; Kant/Reinhardt, 100 Jahre; Krafft, Im Schatten; Weiss, Beschleunigerlaboratorium; Weiss, The "Minerva" Project.

11 Kant/Reinhardt, 100 Jahre.

for a new geological era. Today, the MPIC plays a major role at the national as well as international level in the exploration of questions relating to environmental chemistry, focusing in particular on the nature, mutual influences and characteristics of the bio-, geo-, atmo-, and anthroposphere. Even a first glimpse at the description of activities taking place within the independent departments located there today clearly demonstrates that the current structure of the institute is based on a principle of complementarity. The common thread that permeates these research activities is that the current departments all deal with cycle's processes present in and between the Earth's spheres, with a particular focus on areas of atmospheric chemistry: the research emphases of the Atmospheric Chemistry Department, headed by Johannes Lelieveld (born 1955), include the development and construction of measuring instruments that detect trace gases in the atmosphere, the identification of photochemical reaction chains, as well the development of model-based computer simulations that describe chemical and meteorological processes.[12] The Particle Chemistry Department (formerly the Department of Cloud Physics and Chemistry) is led by Stephan Borrmann (born 1959) and addresses the chemical constitution and physical characteristics of atmospheric aerosol and cloud particles.[13] The third department established in 2012 and headed by Ulrich Pöschl (born 1969), the Multiphase Chemistry Department, highlights research into chemical reactions and the transport and transformation processes between solids, liquids and gases.[14] Finally, in 2015 the institute established the Climate Geochemistry Department headed by Gerald H. Haug (born 1968), where re-

---

12 See MPIC website: <https://www.mpic.de/forschung/atmosphaerenchemie.html>, status: May 23, 2018.

13 See ibid., URL: <https://www.mpic.de/forschung/partikelchemie.html>, status: May 23, 2018.

14 See ibid., URL: <https://www.mpic.de/forschung/multiphasenchemie.html>, status: May 23, 2018.

searchers are investigating the interaction of processes that take place between the Earth System's elements: climate, ocean and atmosphere from short annual periods through times of geological relevance.[15]

The foundation for this current structure was laid in 1968 with the appointment of meteorologist Christian Junge (1912-1996) and the associated establishment of atmospheric chemistry at the institute. Even as late as the mid-1960s, the MPIC's topics were far from research questions about the atmosphere or even about Earth Systems. The integration of atmospheric chemistry in 1968 is thus a key event in the history of the institute as well as the history of Earth System Sciences in the MPG in general. It resulted in extensive restructuring lasting for several decades at the MPIC and laid a cornerstone for the development of a scientific field, which has shaped the overall profile of the MPG until today. The detailed history behind this reorganization is the subject of the first part of this article. The development of the MPIC is closely tied to the gradual expansion of atmospheric sciences throughout the FRG since the end of the 1960s and the ultimate development of this field into an independent branch of research in the first half of the 1980s.[16]

Interdisciplinary scientific research has recognized the atmospheric sciences collectively as a highly relevant subject for research for some time. Numerous studies have traced how topics in the atmospheric sciences have made a decisive contribution to scientific, political and public discourse over the past forty years.[17] Furthermore, a majority of the existing literature has considered the development of the atmospheric sciences as an independent science sector since the second half of the 20th century as an international phenomenon, and has focused

15 See ibid., URL: <https://www.mpic.de/forschung/klimageochemie.html>, status: May 23, 2018.

16 See Schützenmeister: Zwischen Problemorientierung, 109 f.

17 See Böschen, Risikogenese; Grundmann, Transnational Environmental Policy; Conway/Oreskes, Merchants.

on the structures and roles of relevant organizations such as the "Intergovernmental Panel on Climate Change" (IPCC) or the "Global Change Research Program".[18] Moreover, a number of historical and philosophical studies of the instruments and methodological repertoire as well as the credibility of atmospheric research have been conducted, above all focusing on climate research.[19] However, much less attention has been placed on the history of organizations and institutions; when attention has been given, there is often particular emphasis on the role of the U. S.[20] The historical appraisal of the atmospheric sciences in the FRG remains largely unexplored, with a few individual exceptions such as studies on the history of meteorological services[21] and the Institute for Physics of the Atmosphere in Oberpfaffenhofen.[22] The MPIC in Mainz so far has remained largely unnoticed in this context, but nevertheless is a particularly suitable subject for research in order to bring to light further information on the development of research on both the atmosphere as well as the Earth System as a whole. The three chapters of this volume deal, chronologically, with the episodes of the institute's history since the late 1950s.

The first part, an internal perspective of the MPG, traces the search for a successor for former MPIC director Josef Mattauch (1895–1976) that began at the end of the 1950s with a number of failed attempts.

18 E. g. Kwa, Local Ecologies; Beck, Moving; Leuschner, Glaubwürdigkeit; Uhrqvist/Linnér, Narratives of the Past.

19 Amy Dahan (2013): Historic Overview of Climate Framing, in: HAL Working-papers, URL: <https://halshs.archives-ouvertes.fr/halshs-00855311/document>, June 6, 2018.—Edwards, History; Heymann, Understanding; Heymann, Lumping; Gramelsberger, Conceiving process. See also: Gramelsberger/Feichter, Climate Change.

20 Incl. Hart/Victor, Scientific Elites.—Spencer Weart (2010): The Discovery of Global Warming, URL: <https://history.aip.org/climate/>, October 9, 2018.

21 Wege, Entwicklung.

22 See Achermann, Institutionelle Identität.—See also: Volkert/Achermann, Roots; Achermann, Eroberung.

This search plunged the Chemistry, Physics and Technology Section of the MPG (CPT section) responsible for appointments and the institute itself into a severe crisis in the mid-1960s, but, with numerous detours, finally led to the appointment of meteorologist Christian Junge and the establishment of the Departments for Atmospheric Chemistry and Cosmochemistry. The latter was headed by geochemist and meteorite scientist Heinrich Wänke.

Part two is based on an article[23] published in 2016 and looks at the history of the institute under the leadership of Junge until the end of the 1970s, with particular regard to the role of the Atmospheric Chemistry Department and the DFG Special Research Project 73 "Atmospheric Trace Gases" (SFB 73) which largely characterized that department in the context of the development of the atmospheric sciences within the Federal Republic of Germany. The SFB was the first comprehensive DFG program to address the chemical and physical nature of the Earth's atmosphere in the context of which a new generation of young scientists were specifically trained primarily in the study of the chemical composition of the global atmosphere. Through the 1970s, environmental sciences in general, and thus atmospheric research as a part of this field, became increasingly relevant for political authorities as well. Furthermore, the sense of the crucial importance of anthropogenic influences on material cycles started to gain a foothold in research, continuing to mature into a scientific understanding by the middle of the decade.

Finally, the third part of this book concentrates on the continuation of atmospheric chemistry at the MPIC and the increasing focus on questions relating to Earth Systems under Paul Crutzen (born 1933) and Meinrat O. Andreae (born 1949). This part is divided into two sections. The first section, starting with the late 1970s, addresses the further development of the institute into one oriented on the chemistry of the Earth's systems. The focus of this section is both on central

23 Lax, Aufbau.

research topics (in particular the "CLAW" hypothesis and biomass combustion) and the structural reorganization of the institute at the end of the 1970s as well as the establishment of the Biogeochemistry Department under the leadership of Meinrat O. Andreae in 1987. The second section pursues separately Paul Crutzen's work on anthropological influences on the Earth's climate and the Earth System from the start of his scientific career in the early 1970s (the role of aviation, CFCs), through his time at the MPIC (nuclear winter, the ozone hole) and up to the concepts of the Anthropocene and geoengineering that were significantly shaped by Crutzen after his retirement in 2000.

1

INSTITUTIONAL CHANGE BETWEEN TRADITION AND INNOVATION: THE MPIC 1959 – 1968

The Max Planck Society (MPG) is one of the largest and oldest non-university research organizations in the German science landscape. Its vision is primarily characterized by the claim to develop and examine new research areas in fundamental sciences. At the same time, the MPG attaches great value in the emphasis of its own traditions and long-standing institutional structures. The establishment of new research departments at the Max Planck Institutes is usually directly associated with new appointments, at which time the structure of the respective institution is then reorganized or rearranged. In part, this takes place with a delicate balance of contradictions, because, with the exception of start-ups, there are existing institute structures that must be considered when new appointments are involved. At the same time, however, following the Harnack Principle[24]—equally well-known and

24 The Harnack Principle refers to a statement by the first president of the KWG, Adolf von Harnack, who recommended that research institutes must be built around their director as an outstanding scientist. Some points of criticism were based on a single

controversial from the point of view of the history of sciences—top executives from new research fields should be brought to the institutes. They should be given as much individual freedom as possible for shaping their research field. The MPG remains reserved in the disclosure of the decision-making processes that ultimately result in successful appointments and institutional restructurings. From the point of view of (historical) science studies, there is considerable interest in a better understanding of the processes of development and establishment of new research institutions in scientific organizations, in particular if the latter enjoy such high prominence as do those of the MPG. Important questions in this context are: Is each institutional development bound to a path? At what point does it break away from existing situations and promote changes? How do these processes take place?

The 1960s thus offer themselves as a period for investigation from the perspective of the institute as well as from a historical perspective. During this time, the MPG as a whole was undergoing a process of change that could also be seen in the context of the social upheavals of the decade. For the special case of the MPIC, this period offers what is possibly a unique opportunity to trace the course of a reorientation of personnel and thematic focus as well as restructuring taking place over nearly a decade at one of the oldest and most successful Max Planck institutes. It was here that at the end of the 1950s, the search for a successor for then incumbent institute director Josef Mattauch

---

phrasing from Harnack being elevated to form *the* Harnack Principle, even though there are a number of further statements with maxim character. In addition, it is questionable whether the formulation known as the Harnack Principle has applied as a consistent scientific political guideline of the MPG or of its predecessor organization the Kaiser Wilhelm Society (KWG) respectively (see Laitko, Harnack-Prinzip). However, it is beyond doubt that it has served and still does serve a purpose of self-identification in a number of contexts, which is shown by more recent statements by higher MPG representatives (see e. g.: Lüst, Antriebsmotor). This, however, was certainly the case for the 1960s under the presidency of Adolf Butenandt, which is the focus of this article (see Laitko, Harnack-Prinzip, 164 ff.).

began which was the start of a veritable odyssey of failed appointment attempts and re-orientations in thematic focus. In 1968, after closing the institution had been intermittently discussed, a successful conclusion was reached that was surprising and would turn out in equal measure to characterize the institute in the following decades. The MPIC, which under Mattauch had focused primarily on radio- and physical chemistry at the beginning of the 1960s, now had Christian Junge, who was the first director with a meteorological focus. This appointment involved extensive changes to the institution as well as to technical aspects of operations. Thus, the tradition of a patriarchy-dominated institute management by a single director that had been upheld until that point in the MPG was broken, and he was replaced by a board of directors that consisted of the department heads at the institute. In addition, a department for "Chemistry of the Atmosphere and Physical Chemistry of Isotopes" was established, laying the foundation for the present structure of the institute. Up to the present, the MPIC has developed into an institute for "Earth System Sciences" that is highly regarded around the world and that focuses on investigating the Earth's atmosphere and its mutual relationships with the geo-, bio-, and hydrospheres.[25] These developments were unforeseeable at the beginning of the 1960s; at that point, it was difficult to find any candidate wishing to serve as director of the MPIC. In 1963, Werner Heisenberg, Nobel Prize winner and director at the MPI for Physics in Munich, presciently predicted after the first failed MPIC appointments that it would probably be

> "the same again as in the last 15 years, that one sees: the aspect of continuation of the institute is contrary to the other aspect of giving the M. P. G. a new face."[26]

25 For this see also the research overview of the MPIC website: <https://www.mpic.de/forschung/uebersicht.html>, status: May 23, 2018.

26 Minutes of the appointment committee meeting at the MPIC on March 1, 1963, in: AMPG, II. Abt., Rep. 62, No. 494.

As a case study, the MPIC offers the opportunity to gain and visualize deep insights into the methods for appointments and the processes of structural change at the MPG during the 1960s and to see how the contradiction between innovation and tradition pointed out by Heisenberg appeared in such contexts, or if it could be resolved. This clearly shows that, after almost a decade full of setbacks, the strategies for appointments gradually changed, and became somewhat more flexible, and it became possible to establish a novel research field in the Federal Republic of Germany.

Initially, three general phases can be distinguished. The first phase can be described as the "traditional phase". It began in 1960 and ended in 1963. At this time, the potential for innovation was at the lowest point. During this period, appointments were primarily aimed at obtaining a director who could immediately meld with the existing institute structures and, in particular, with Josef Mattauch's thematic priorities. Phase two is characterized by the wave of appointments that followed, in which appointments were gradually and unavoidably freed from pure thematic compatibility. Increasingly, the ability of the candidates to create an association was no longer defined at a thematic level but was instead more aligned with methodology. Ultimately, this change prepared for phase three and made it possible to establish a research field that before then had never been cultivated: research of the chemistry of the Earth's global atmosphere. Phase three is marked by a time of crisis, in which closure of the institute was one option that was discussed and that ultimately concluded with the appointment of Christian Junge. These three phases are clearly reflected in the names of the committees responsible for the appointments in each case. At the beginning, the "Paneth Successor" committee was established, which in the years that followed was initially changed to "Appointments to the MPIC" and finally to "Future of the MPIC".

In the following, we will take a look at the initial situation at the institute at the end of the 1950s.

1.1 Initial situation at the end of the 1950s

At the end of the 1950s, under the leadership of Josef Mattauch, the fields of work were divided into four departments, of which physical chemistry, as the director's own focus of interest, was clearly the dominant one. In addition to the Department for Mass Spectroscopy headed by Mattauch himself and focusing on the binding energies of atomic nuclei, there was a Department for Mass Spectroscopy II under Mattauch's former student Heinrich Hintenberger, with a focus on isotope cosmochemistry.[27] The other research centers were a Department for Nuclear Chemistry headed by physicist Hermann Wäffler and finally the Department of Radiochemistry led by the institute's second director, Friedrich A. Paneth, whose influence, however, paled in comparison to that of Mattauch. Paneth's main focus was the measurement of noble gases in stone and iron meteorites. He came to the MPIC as successor to the co-discoverer of nuclear fission, Fritz Straßmann, but died in 1958 and thus his position became vacant at an early stage.[28] His work, nevertheless, laid the cornerstone of the second strong research field at the institute: cosmochemistry.

As Mattauch's retirement scheduled for 1963 neared, the plan for a successor to the directorship gradually became more pressing. Considering this initial situation, and the name of the Max Planck Institute *for Chemistry*, the ultimate establishment of atmospheric scientific research that was just at the beginning at this time in the FRG did not appear to be an obvious matter of course. An episode lasting almost a decade and characterized by failures and coincidences was necessary for this step. First of all, a successor for Friedrich Paneth needed to be found, who could lead the institute after Mattauch's retirement;

27 Hahn sent Hintenberger a letter on June 15, 1959 to confirm that Hintenberger was henceforth to be director of the MPIC and head of an independent department: Hahn to Hintenberger, on June 15, 1959, in: AMPG, III. Abt., ZA 95, folder 2.

28 See Kant/Lax, Chronik, 271.

thus the first committee (“Paneth Successor”) was established to seek out potential candidates. The commission consisted of members of the Chemistry Physics and Technology Section (CPT section) of the MPG which already included a core group of players who would dominate the discussions on suitable candidates in the coming years. In addition to the former director of the MPIC already mentioned, Josef Mattauch, and Werner Heisenberg, particularly active members of this group included physical chemist and former co-worker of Wernher von Braun (1912 – 1977) Carl W. Wagner (1901 – 1977), director of the MPI for Physical Chemistry until the mid-1960s, who initially presided over the CPT section during that period. Werner Köster (1896 – 1989, director of the MPI for Metal Research) took over the section as well as the leadership of the future committee “Future of the MPIC”. Director of the MPI for Physics, Wolfgang Gentner (1906 – 1980), who was also well-networked with the Gutenberg University in Mainz, also played an important role. Otto Hahn (1879 – 1968), former director of the MPIC and president of the MPG until 1960, was also continuously involved in the discussions, though less assertively than would have been expected considering his connection to the MPIC that since 1959 had borne the name affix “Otto Hahn Institute”. Outside the appointment committee, players of the MPIC Board of Trustees were also involved, in particular Hoechst Chairman Karl Winnacker (1903 – 1989) and Otto Hahn’s successor as MPG president Adolf Butenandt (1903 – 1995). Of course, it should be noted that the influence of the respective players was not constant over the years and the discussions shifted here and there to some extent.

As early as 1960, the “Paneth Successor” committee initiated the first serious appointment attempt with an effort to win Hans Eduard Suess (1909 – 1993) for the institute. In focusing on geochemistry, Suess’ research interests in physical chemistry meant that his focus was aligned with that of the incumbent MPIC director Josef Mattauch[29].

29 See Kant et al., Wissenschaftliche Mitglieder, 357.

Fig. 1: Josef Mattauch, Heinrich Hintenberger, Hermann Wäffler, Friedrich Paneth (from top left to bottom right)

He had received his qualification as a professor in Hamburg and had worked on the German Uranium Project in Hamburg during the Second World War, where he worked primarily on the production of deuterium oxide (heavy water). In the early 1950s, he went to the University of Chicago to work with the group of Harold Urey (1893 – 1981), who himself was the discoverer of deuterium.[30] For Suess' appointment, it is quite possible that the motif of repatriation of researchers who established themselves outside of the FRG during and after the Second World War played an important role. However, there is no clear evidence regarding this in the files of the institute and the board of trustees. But as we will see later, at times, this point played a central role for other candidates.

Suess' research interests included the analysis of cosmic frequency distributions of elements and hence he also worked on the radioactivity of potassium and its use for determining the age of elements in meteorites.[31] This application would have provided a direct connection to those research works that were carried out before in Paneth's Department for Age Determination of Iron Meteorites. The combination of his fields of work made Suess an ideal candidate, since he covered Mattauch's previous core area as well as the geochemical work by the "Meteorite Research Group" that remained at the institute. Therefore, it is no surprise that every effort was made to offer the best possible conditions for Suess and his research at the MPIC. Indeed, in 1960, having in the meantime qualified as a professor at the University of California in San Diego,[32] he came to the institute in Mainz for just under a year. There, he first worked representatively in Paneth's position and it was during this stay that an X-ray diffractometer was

30 For the discovery of heavy hydrogen (deuterium) in 1931, Urey received the Nobel Prize for Chemistry in 1934. See Harold Urey, URL: <https://www.nobelprize.org/prizes/chemistry/1934/summary/>, status: October 9, 2018.

31 See Wänke/Arnold, Hans E. Suess, 4.

32 See ibid., 4.

purchased for 73,000 DM specifically for him.[33] After this period, however, Suess decided against the position offered to him and on January 31, 1961, he informed the president of the MPG, Adolf Butenandt, by letter of his decision to return to San Diego.[34] Presumably, a significant reason for his decision was that Paneth's successor was initially not linked to the position of chief director of the institute—which was still held by Mattauch for the time being. This conclusion draws from the fact that, some years later, in the framework of an informal conversation, Suess reacted very positively when asked if he could imagine following Mattauch as chief director of the institute.[35]

It is difficult to find a definitive answer to the question of how much Suess' rejection in 1961 was a significant factor in the difficulty the MPG encountered in promoting the Meteorites Group at the MPIC over the next several years. Nevertheless, after Suess' rejection, Otto Hahn, MPIC grandee and co-discoverer of nuclear fission, explicitly stated to the board of trustees that *"even if there were a change in the chemistry department, the specific meteorite research working group will want to remain at the institute."*[36] What is certain, however, is that Paneth's former employees were no longer primarily financed by the MPG from this year on but instead received third-party funds provided by the Fritz Thyssen Foundation.[37] Between 1961 and 1964, these funds amounted to a sum of close to 1,083,100 DM;[38] this source

33 Seeliger to Mattauch, May 24, 1960, in: AMPG, III. Abt., ZA 95, folder 2.

34 See Suess to Butenandt, January 31, 1961, in: AMPG, II. Abt., Rep. 66, No. 842.

35 See Gentner and Köster, April 27, 1966, in: AMPG, II. Abt., Rep. 62, No. 495.

36 Minutes of the board meeting of the MPIC, February 17, 1961, p. 4, in: AMPG, II. Abt., Rep. 66, No. 849.

37 See Carsten Reinhardt/Gregor Lax: Interview with Friedrich Begemann, January 6, 2012, in Mainz.

38 This sum is derived from notifications on annual grants. See: Ballreich an Coenen, October 20, 1961, in: AMPG, II. Abt., Rep. 66, No. 842.—See also: Coenen to Ballreich, February 5, 1962, in: AMPG, II. Abt., Rep. 66, No. 843.—Seeliger to Mattauch, November 21, 1961, in: AMPG, II. Abt., Rep. 66, No. 842.—See also:

of support disappeared in 1965. At times during the beginning of the funding period in 1962, significant sums of the now available MPIC funds were earmarked for other institute departments; specifically, the Department for Mass Spectroscopy headed by Heinrich Hintenberger profited in particular with a sum of 268,000 DM and, to a much lesser extent, Hermann Wäffler's Nuclear Physics Group also benefited with 35,000 DM.[39] But just when it began to appear in the same year that the succession question would continue to remain uncertain, the institute's board of trustees recommended saving for the time being the MPIC funds that had been freed up until 1964 by the funds from the Fritz Thyssen Foundation.[40] In the meantime, the Meteorites Research Group itself had been integrated into Hintenberger's Department and received its funding there once the Thyssen support stopped and up until the recovery of its autonomy in the form of the Department for Cosmochemistry that was later established.[41]

After Suess' rejection, several candidates were contacted, amongst others the physical chemist Paul Harteck (1902 – 1985), who had already been discussed in 1953 as successor to Paneth's predecessor Fritz Straßmann[42] and who had worked until 1933 as assistant to Fritz Haber (1868 – 1934) at the KWI for Physical Chemistry and Electro-

Seeliger to Mattauch, November 28, 1961 and Ballreich to Mattauch, November 2, 1961, in: AMPG, III. Abt., ZA 95, folder 2.—Domstreich to Roeske, April 13, 1964, in: AMPG, II. Abt., Rep. 66, No. 840—Scholz to Coenen, January 11, 1965, in: AMPG, II. Abt., Rep. 66, No. 842.—Coenen to the MPG General Administration, March 7, 1964.

39 See note Roeske, reg. MPI Chemistry—Meteorites Group, Göttingen, October 2, 1962, in: AMPG, II. Abt., Rep. 66, No. 842.

40 See note Roeske, reg. Chemie Mainz—Continuous Budget 1963, October 29, 1963, in: AMPG, II. Abt., Rep. 66, No. 842.

41 This was already decided in March 1964. See note Roeske, March 10, 1964, p. 1, in: AMPG, II. Abt., Rep. 66, No. 843.

42 See Hahn to Butenandt, January 9, 1961, in: AMPG, II. Abt., Rep. 66, No. 840.

chemistry in Berlin-Dahlem.[43] Harteck played a key role in the German Uranium Project and was therefore interned by the Allies at Farm Hall in 1945.[44] However, he rejected the offer in advance, pointing out that he would choose a position in Munich over one at the institute.[45] It was not until the end of 1961 that serious negotiations were carried out with physicist Sven Gösta Rudstam (1925 – ?), who headed the newly established research group for nuclear chemistry at the Centre Européen de Recherche Nucléaire (CERN).[46] Initially, it was considered whether Rudstam, or even better, his previous teacher Alexis C. Pappas (1915 – 2010) should be appointed, but ultimately a decision was made for the former. The reasons for this were both that Rudstam could have continued his working focus "after installation of the linear accelerator intended for Mainz" and also that the likely rejection of Pappas was anticipated since he was firmly established at Oslo University, where a new laboratory had just been built especially for him.[47] The linear accelerator in Mainz was a joint project equally supported by the University of Mainz and the MPIC. Although not put into operation until 1967, the relevant contract on use and financing had already been drafted between the two institutions in 1958 and was only slightly modified in the years that followed.[48] The negotiations

43 See Klee, Personenlexikon, 228.

44 See Michael Schaaf: Schweres Wasser und Zentrifugen. Paul Harteck in Hamburg (1934 – 1951), URL: <http://censis.informatik.uni-hamburg.de/publications/Art_M_Schaaf_Harteck.pdf>, October 1, 2014, p. 7.—See also the comprehensive science historical biography about Harteck: Schaaf, Physikochemiker.

45 See Hahn to Butenandt, January 9, 1961, in: AMPG, II. Abt., Rep. 66, No. 840.

46 See Sven Gösta Rudstam: Leader of the Nuclear Chemistry Group in the Nuclear Physics Division, in: Who's who in CERN, URL: <http://lib-docs.web.cern.ch/lib-docs/Archives/biographies/Rudstam_G-196303.pdf>, status: May 23, 2018.

47 Meeting of the "Paneth Successor" Group, October 5, 1961, in: AMPG, II. Abt., Rep. 62, No. 494.

48 Contract between MPIC and the University of Mainz on the use of the linear accelerator, November 25, 1958, and amendment agreement February 16, 1961, in: AMPG, II. Abt., Rep. 66, No. 844.

with Rudstam were difficult from the beginning. His excellent position at CERN necessitated considerable concessions in salary classification on the part of the MPG. The typical highest pay group of an MPG director at the time, which paid an annual salary of 25,931.64 DM, was far from sufficient.[49] "*We must offer Prof. Rudstam an allowance of 8,000 DM and a duty allowance of 1,200 DM to even come close to his present salary at the CERN in any way [sic!].*"[50] Mattauch exerted his influence together with physicist Wolfgang Gentner, who was also a member of the "Paneth Successor" committee and had also worked at CERN.[51] Both strongly advocated the appointment of Rudstam and left no doubt about his suitability to significantly enrich the institute.[52] The negotiations very soon started to drag on, since Rudstam had delayed talks on an offer from Sweden at the same time.[53] It took until the end of 1962, before the rejection was finally received by the president of the MPG.[54] Meanwhile, there was serious pressure to act, since Mattauch's retirement was already imminent in the following year. As a consequence, the "Paneth Successor" committee was replaced with the "Appointments to the Max Planck Institute for Chemistry" committee in the first quarter of 1963. This committee again consisted of the hard core listed earlier (and a number of other players from the CPT section who in part, however, attended the meetings only sporadically) and was supplemented with "guests"

49 See in this regard the itemized list of compensations of department heads at the MPIC, January 25, 1962, in: AMPG, II. Abt., Rep. 66, No. 843.

50 Note Seeliger, January 26, 1962, in: AMPG, II. Abt., Rep. 66, No. 843.

51 See Gentner's scientific achievements in detail: Ulrich Schmidt-Rohr (2008): Wolfgang Gentner. 1906 – 1980, URL: <http://www.uni-frankfurt.de/fb/fb13/Dateien/paf/paf181.html>, status: May 23, 2018.—See also the commemorative publication published previously: Hoffmann/Schmidt-Rohr, Wolfgang Gentner.

52 See Mattauch to Butenandt, May 28, 1962, in: AMPG, II. Abt., Rep. 66, No. 843.

53 Rudstam to Mattauch, May 23, 1962 and Mattauch to Butenandt, May 28, 1962, in: AMPG, II. Abt., Rep. 66, No. 843.

54 Rudstam to Butenandt, November 16, 1962, in: AMPG, II. Abt., Rep. 66, No. 843.

and then reduced again in 1964. Attendees exclusively included other MPG directors such as Ludwig Biermann (1907 – 1986, MPI for Physics and Astrophysics), Rudolf Brill (1899 – 1989, MPI for Physical Chemistry and Electrochemistry), Ulrich Schmidt-Rohr (1926 – 2006, MPI for Nuclear Physics) and Karl Ziegler (1898 – 1973, MPI for Coal Research).[55] Heinrich Hintenberger from the MPIC and Otto Hahn were frequently present as guests in following years.

In light of the rejections from Suess and Rudstam, which had been accepted at that point, they first found themselves facing the need to retire Josef Mattauch (then 68 years old), two years later than originally intended, and to extend his contract until 1965. People at the institute were not especially happy with this decision and Mattauch himself didn't start his period of extension without explicitly expressing his concerns on this matter.[56]

Fears about the fate of the institute arose not only among the scientists but also within the board of trustees of the MPIC. The board consisted firstly of scientific members from the institute, secondly of representatives from the MPG general administration as an umbrella institution, and thirdly of a number of members who could to some extent be better assigned to political and/or economic sectors than to (basic) sciences. Although the board of trustees itself had no official mandate to intervene in institute affairs, they naturally attempted to influence the future of the institution, specifically from the economic side. In particular, this included the influential Karl Winnacker who was chairman of Farbwerke Hoechst AG already at the beginning of the 1950s, notwithstanding his questionable career at the I. G. Farben

55 See minutes of the "Appointments to the MPIC" committee meeting on March 1, 1963, in: AMPG, II. Abt., Rep. 66, No. 840.

56 See Mattauch to Butenandt, March 26, 1963, in: AMPG, II. Abt., Rep. 25, (A50).

under the Nazi regime.[57] As president of the MPIC board of trustees,[58] Winnacker got involved in the question of the future continuation of the institute at an early stage and, in 1963, expressed his concern to MPG president Butenandt that after the chemistry group was "paralyzed" by the death of Paneth, and now, with Mattauch's retirement, the end of physical chemistry at the institute would also lie ahead.[59]

Although, at least at the level of the academic members, the physicist Hermann Kümmel (1922 – 2012) was appointed in 1963, a tendency towards general uncertainty even for the appointments for other posts at the MPIC is evident at this time. A particularly interesting example is the case of Heinrich Wänke, who, with the appointment of Christian Junge, was ultimately also to take over leadership of an independent department as director at the MPIC in 1968.

Josef Mattauch had proposed back in 1963 to appoint Wänke as scientific member of the institute, but met once again with strong opposition. Winnacker's description of the state as "paralysis" was thus in some aspects very fitting.

Direct criticism of institute director Josef Mattauch was growing in the committee and within the entire CPT section and culminated in a letter from section chairman Carl Wagner sent by courier in March 1963. The letter was written in consultation with Otto Hahn and Wagner accused Mattauch of significant involvement in the unsatisfactory situation at the MPIC due to previous poor decisions about personnel and denied him de facto the competence for further pro-

57 For Winnacker's role at I. G. Farben and his SA and NSDAP past, see: Lindner, Hoechst, 211 ff.

58 Winnacker was initially deputy chairman of the board of trustees in 1955 and took on the position as chair some time later, a post he held until the 1970s. See the lists of members of the board of trustees, in particular on August 23, 1955 and May 1, 1959, both in: AMPG, II. Abt., Rep. 66, No. 848.—See also the list of the board of trustees in the minutes of the MPIC meeting on December 5, 1973 in: AMPG, II. Abt., Rep. 66, No. 851.

59 Winnacker to Butenandt, January 28, 1963, in: AMPG, II. Abt., Rep. 66, No. 850.

posals.[60] Wagner considered his accusations to be confirmed by the fact that Mattauch had proposed Hermann Wieland for Paneth's position two years ago, although, in Wagner's eyes, Wieland's research focuses were not reasonably compatible with the departments at the institute at that time.[61] Thematic connectivity still had priority as a selection criterion at this time. In 1961, Wieland had been seriously discussed by the "Paneth Successor" committee as a possible candidate. His main focuses were organic chemistry and there was an initial thought about a connection to roots that were long standing but no longer up-to-date, with particular attention to the works of the Nobel Prize winner Richard Willstätter in the mid-1910s.[62] In addition to Wieland, other organic chemists were discussed, but the idea to further develop organic chemistry in Mainz was subject to fundamental criticism. Making reference to various other MPIs working on different questions and ranges of topics in organic chemistry[63] ultimately proved effective against such a "rehabilitation". Inorganic chemistry was also considered for a short time, although this idea was quickly abandoned. This topic area was also not close enough to the fields of the existing departments. Still in 1964, there was a warning of the potential for a new appointment to create excessive heterogeneity within the institute structure.[64] This idea was entirely rejected by the committee, in particular by Werner Heisenberg and specifically with the following

60 See: Wagner to Mattauch, March 6, 1963, Bl. 1, AMPG, III. Abt., ZA 95, folder 2.

61 Ibid.

62 See: Minutes of the "Paneth Successor" committee meeting on February 25, 1961, Bl. 2 f., in: AMPG, II. Abt., Rep. 62, No. 494—Willstätter ran the department for organic chemistry at the KWIC between 1912 and 1916 (cf. Kant et al., Wissenschaftliche Mitglieder, 366 f.).

63 See: Minutes of the "Paneth Successor" committee meeting on February 25, 1961, ibid.

64 In the documents of the "Appointments to the MPIC" committee dated June 8,1964, reference is made in this regard to a circular for the committee members dated January 13, 1964, Bl. 3, in: AMPG, II. Abt., Rep. 62, No. 494.

reference: if a new foundation *"took place with any new appointment, the MGP would reach its end"*.[65]

It became apparent with regard to a possible establishment of organic and inorganic chemistry, that the responsible committee members were gradually losing inspiration for any "realignment". On top of this, in the case of inorganic chemistry, institute traditions from the episode of the German Empire were being used as a legitimation strategy for the appointment of new candidates for the director posts. Furthermore, it is evident that in the early 1960s there was a desire for the future institute alignment, in terms of its range of topics, to not stray too far from the institute's existing areas. This had previously been clearly reflected in the profiles of the first two appointment candidates, Suess and Rudstam. Both were directly compatible with the institute, in particular with physical chemistry in line with Mattauch that up to then had been firmly anchored to the institute. For Suess there was also a significant bridge to meteorite research.

Ultimately, in the phase that followed, there was a move away from the requirement that candidates be predominantly compatible in terms of topics, and movement towards compatibility with the methods anchored in the institute. This was what made it possible to speak about the integration of disciplines that were located outside the topic areas previously worked on at the MPIC. The first candidate appointed by the subsequently newly formed "Appointments to the MPIC" committee in late 1963 was the physicist and Nobel laureate Rudolf L. Mößbauer (1929 – 2011).

65 Minutes of the "Appointments to the MPIC" committee meeting, May 13, 1963, Bl. 11, in: AMPG, II. Abt., Rep. 62, No. 494.

1.2 The "Appointments to the MPIC" Committee

In December 1963, the CPT section decided to offer an appointment as director of the MPIC to Rudolf Mößbauer.[66] Like Rudstam, Mößbauer's focus was much less on chemistry and more in the realm of physics. In 1957, he did his PhD thesis at the Institute for Physics of the MPI for Medical Research in Heidelberg. His supervisor was Heinz Maier-Leibnitz. At the time of the invitation of the MPIC, he was professor of physics at the California Institute of Technology (CalTech) in Pasadena.[67] A connection between his research activities and the work being conducted up to that time at the MPIC was based not so much on a specific topic, but rather on tools and methodological innovation. His work on resonance absorption of gamma radiation that resulted in the discovery of the aptly named "Mößbauer effect"[68] made it possible to significantly improve the measurement accuracy of mass spectroscopic examination methods. For these works, Mößbauer received the Nobel Prize in Physics together with Robert Hofstadter in 1961.[69]

The idea to appoint Mößbauer to the MPI was very positively accepted by the committee with a ratio of ten votes to one.[70] The desire to lure a highly renowned physicist who had migrated to the US back

66 Minutes of the meeting of the Chemistry, Physics and Technology Section in Frankfurt am Main (extract), December 4, 1963, in: AMPG, II. Abt., Rep. 66, No. 841.

67 See Rudolf Mößbauer, Kurzbiographie, URL: <https://www.nobelprize.org/prizes/physics/1961/mossbauer/biographical/>, status: October 9, 2018.

68 The Mößbauer effect refers to the phenomenon that emission or absorption of a gamma quantum passes through an atomic nucleus without recoil when the nucleus is surrounded by a crystal lattice that can take over the rebound without significant extraction of energy from the gamma quantum. An introduction on the Mößbauer effect can be found in: Lustig, Mössbauer Effect.

69 See Kazemi, Nobelpreisträger, 219.

70 See minute of the "Appointments to the MPIC" committee May 13, 1963, p. 16, in: AMPG, II. Abt., Rep. 62, No. 494.

to the FRG clearly played a role in this as well.[71] Under the purport of this "repatriation", Mößbauer's appointment became a matter of public interest when reported in a short article in the "Spiegel".[72] Although the desire for "repatriation" was fulfilled in general, the hope to acquire the Nobel Prize winner for the MPIC was shattered. At the beginning of 1963, Mößbauer's written cancellation reached Butenandt. The reasons given were that the MPG's freedom to choose not to divide the institute should be restored[73] and that, in any case, Mößbauer wanted to wait for a concrete appointment offer from his clear favorite, TH Munich.[74] In the following year, he accepted a professorship for experimental physics there and became director of their newly established physics department, replacing his former teacher Maier-Leibnitz, who had in the meantime relocated to the TH.[75]

In the period that followed, numerous potential proposals were discussed and rejected by the appointment committee. Some of these candidates also came from physics and not from chemistry, including Arnold Schoch (1911 – 1967) from CERN and Aaldert Hendrik Wapstra (1922 – 2006) from the Amsterdam Instituut voor Kernfysisch Onderzoek; the latter had also already collaborated successfully with Josef Mattauch for some time.[76] Slowly but surely, however, the list of possible candidates who were considered on the basis of their research focuses ran out and there was a gradual move to be more flexible about

71 The "Brain-Drain" that involved targeted headhunting of elites from Germany after the Second World War by the allied occupying powers was widely noticed with major worries. Particularly during the 1950s, a real residue topos concerning research and development arose in the young FRG. See therefore also Lax, Lineares Modell, 71.

72 See Rückführung, 12.

73 There had earlier been thoughts about outsourcing parts of the institute.

74 Mößbauer to Butenandt, December 2, 1963, in: AMPG, II. Abt., Rep. 62, No. 495.

75 See Kazemi, Nobelpreisträger, 219 ff.

76 See meeting of the "Appointments to the MPIC" committee June 8, 1964, in: AMPG, II. Abt., Rep. 62, No. 494.

potential options. It had finally become a serious notion to strengthen theoretical chemistry within the MPG and therefore fundamentally in the FRG, which was thought to be behind in this field compared to other nations.[77] Thus, the idea to acquire Hermann Hartmann (1914–1984) for the MPIC arose. Hartmann also came from physical chemistry, but unlike the previous candidates, his main focus was on theoretical chemistry. In 1952/1953, he was working at the MPI for Physical Chemistry and Electrochemistry and then accepted a professorship and simultaneous directorship at the Institute for Physical Chemistry at Frankfurt University, and he went on to significantly contribute to the instittue's development in the following years.[78] Of course, there were critical voices within the MPG in Hartmann's case as well, in particular from CPT section chairman Wagner. He addressed the president of the MPG in writing and informed him that he didn't consider Hartmann an ideal candidate. The reason he gave was that Hartmann's appointment would not lead to sufficient innovation of the specialist orientations at the institute, because he

> "[places] great emphasis on not losing the connection to the experiment. The connection between theory and experiment is Hartmann's special strength [...] In conclusion, the appointment of Hartmann at Mainz would be neither an optimal promotion of the field of theoretical chemistry nor an especially favorable solution for the appointment problem in Mainz."[79]

Wagner's criticism was essentially that the distinct focus on pure theoretical chemistry, and hence the establishment of a branch that did not yet exist at the institute, could not be realized radically enough with Hartmann. His proposed solution was to post Hartmann to a branch office for theoretical chemistry in Frankfurt—nevertheless, he

---

77 See ibid.

78 See minutes of the Chemistry, Physics and Technology Section of the MPG, June 21, 1965, pp. 3 f., in: AMPG, II. Abt., Rep. 62, No. 494.

79 Wagner to Butenandt, June 19, 1964, p. 4, in: AMPG, II. Abt., Rep. 62, No. 495.

still didn't feel that this was the best solution.[80] Wagner was not the only critic. As the appointment became closer to a reality, Nobel Prize winner Karl Ziegler from the MPI for Coal Research also expressed criticism, stating that

> "the opinion of capable young gentlemen well-versed in the field of theoretical chemistry and in whose view Hartmann's showcasing is hard to understand … reinforced his [Ziegler's] rejection of Hartmann."[81]

However, this attitude was not shared by the majority, and perhaps it was the very balance between experiment and theory that Hartmann maintained that had piqued the interest of the MPIC in the first place, since until then there had been a strong focus on experimentation.

In addition, Hartmann was the first candidate not recruited from abroad; the motivation for "repatriation" that had clearly been in the foreground in the case of Mößbauer and probably also with Suess had taken a back seat. In particular, MPG president Butenandt got involved in the negotiations with Hartmann. Although Hartmann had made it clear from the outset that his goal was to establish a center for theoretical chemistry and that opportunities for this were available in Frankfurt through sponsorship that had been proposed for several years by the Volkswagen Foundation,[82] the negotiations with the MPG were apparently positive at the beginning. Both Butenandt's own assessments[83] and the fact that the critical Wagner of all people (here in the

80 Ibid., 5 f.

81 Minutes of the "Future of the MPIC" committee meeting, June 21, 1965, p. 4 f., in: AMPG, II. Abt., Rep. 62, No. 494.

82 See Wagner to Butenandt, June 19, 1964, pp. 2 f., in: AMPG, II. Abt., Rep. 62, No. 495.

83 See note Butenandt telephone conversation with Hartmann, July 5, 1965 and Butenandt to Köster, May 28, 1965, in: AMPG, II. Abt., Rep. 66, No. 843.—Butenandt also reported to the senate on the progress in the negotiations with Hartmann. See minutes of the 51st Senate Meeting of the MPG in Ludwigshafen, p. 19, in: AMPG, II. Abt., Rep. 60, No. 51.SP.

function of his presidency of the CPT section) had already written a congratulatory speech for Hartmann's inauguration, were testament to this view.[84] It was of no use: in 1964, Hartmann had already declined an offer from TH Munich in order to stay on in Frankfurt and, in March 1966, the senate of the MPG had to accept that the CPT section once again had to deal with a new appointment.[85]

Furthermore, the integration of the meteorites research group into Hintenberger's department had already shown that employees within the institute were no longer open-minded towards the establishment of new departments. The reason was that the funding for Paneth's former Radiochemical Department now went to Hintenberger's Mass Spectroscopy Department, from where it would have to be withdrawn to finance a new department.[86] The institute structure that was ultimately established later with the Department for Atmospheric Chemistry on the one hand and the Cosmochemistry Department on the other hand was less likely than ever in the mid-1960s.

1.3 Crisis and "Future of the MPIC"

During the negotiations with Hartmann, there was already serious discussion as to how the MPIC could remain a sustainable institution at all. As early as 1964, there was a proposal to transfer the existing departments to other MPIs and phase out the institute. However, there was reluctance to present this idea too blatantly, in part because no one wanted to suffer ignominy in the eyes of the Ministry of Edu-

84 Wagner to Köster, April 27, 1965. [The original is presumably misdated, it should be dated 1966, GL], p. 3, in: AMPG, II. Abt., Rep. 62, No. 495.

85 Protocol of the 53rd MPG Senate Meeting in Hannover, p. 14, in: AMPG, II. Abt., Rep. 60, No. 53.SP.

86 See note Roeske, March 10, 1964, in: AMPG, II. Abt., Rep. 66, No. 843.

cation and Cultural Affairs.[87] Another idea mentioned earlier which was not implemented, but nevertheless pursued, was the reopening of the institution as an Institute for Inorganic Chemistry. However, this most likely would have involved a complete change of location, since Mainz had been deemed relatively unsuitable for such an institute.[88] The general situation at the MPIC was further exacerbated when, in 1965, some months earlier than originally planned, Mattauch could no longer fulfill his duties for health reasons and had in the meantime been represented by Heinrich Hintenberger.[89] Moreover, in 1964 the board of trustees ceased their activities and would not meet again until after the appointment of Christian Junge in 1968.[90] Thus, all authority regarding future decisions lay with the committee and the CPT section. By then, discontent was expressed even in the MPG Senate *"that the [Chemistry, Physics and Technology, GL] section responsible for the appointment of this institute that had such a good reputation and bears the name of the honorary president needs so much time"* to fill the vacant directorship.[91] Under these circumstances, Hartmann's rejection was to result, inevitably, in a serious crisis. In 1965, the responsible committee, now acting under the name "Future of the MPIC", discussed appointment possibilities and other options, to some degree as though starting from scratch. The number of participants was decreased; the players, however, continued to comprise members of the "Appointments to the MPIC" committee: Gentner, Heisenberg, Wagner, Köster, Brill, and Ziegler remained. By then, the MPIC was represented by

87 See Wagner to Ziegler, March 18, 1964, in: AMPG, II. Abt., Rep. 62, No. 495.

88 See Wagner to the members of the Chemistry, Physics and Technology Section, January 13, 1964, in: AMPG, II. Abt., Rep. 62, No. 494.

89 See protocol of the 50th MPG Senate Meeting, March 12, 1965 in Berlin, in: AMPG, II. Abt., Rep. 60, No. 50.SP.

90 See note Roeske, May 16, 1968, board meeting of the MPI for Chemistry in Mainz, May 24, 1968, in: AMPG, II. Abt., Rep. 66, No. 850.

91 Protocol of the 50th MPG Senate Meeting, March 12, 1965 in Berlin, in: AMPG, II. Abt., Rep. 60, No. 50.SP.

Hintenberger and Hermann Wäffler; Mattauch, now retired, as well as Biermann and Hahn, attended the meetings as guests.[92]

Gentner brought Hans Suess back into play; in the meantime, however, this proposal met with clear contradiction in particular from the two MPG grandees Hahn and Heisenberg, but also from Wagner, as *"they felt they had to deny him [Suess], as a real individualist, the suitability as institute director, with all due respect for his professional qualifications."*[93] Werner Heisenberg himself proposed Heinrich Wänke, who had been introduced by Mattauch some years ago but who was deemed by the members of the committee to be too young for such a task.[94] The ensuing negotiations had by then taken an alarming turn for the institute and culminated in the fact that none other than Heisenberg himself was once again proposing to outsource the institution from the MPG and to connect it with Mainz University, under the leadership of Heinrich Wänke.[95] It appears that Heisenberg, out of fear for the good reputation of the MPG, wanted to phase out the institute at any cost, since *"all measures should already be taken to close the institute after the retirement of Hintenberger and Wäffler"* even if the contractual terms of the scientific members at the MPIC needed to be observed.[96] Finally, in 1967, he stated out loud that he would miss fundamental research taking place at the MPIC that *"surpassed an intermediate level"*.[97] Heisenberg's fears for the good reputation of

92 Minutes of the "Future of the Max Planck Institute for Chemistry" committee meeting on June 21, 1965, in: AMPG, II. Abt., Rep. 62, No. 494.

93 Minutes of the "Future of the Max Planck Institute for Chemistry" committee, meeting on May 17, 1966, in: AMPG, II. Abt., Rep. 66, No. 841.

94 Ibid.

95 Minute of the "Future of the MPIC" committee , February 24, 1966, in: AMPG, II. Abt., Rep. 66, No. 841.

96 Minutes of the "Future of the MPIC" committee meeting on April 6, 1967, in: AMPG, II. Abt., Rep. 62, No. 494.

97 Extract from the protocol of the MPG board of directors meeting, June 7, 1967 in Kiel, in: AMPG, II. Abt., Rep. 66, No. 843.

the MPG were fanned by the fact that the difficulties with the appointments had obviously been leaked to the outside. The Scientific Council had listed the MPIC as a negative example for misleading institute continuation, and this criticism did not remain without consequences, as was clearly indicated by a proposal by later MPG president Reimar Lüst to the CPT section. He said that he

> "has rarely heard such harsh negative criticism about one of our institutes; this institute was mentioned as prime example in a discussion of the question whether a focus once accepted should be closed again later."

The objective was now to review "whether the institute could continue under the current conditions".[98]

Although ultimately the fundamental question of a possible dissolution of the institute was pushed to the background, it was apparent that the situation was more than tense. The fact that the name affix "Otto Hahn Institute" finally gave the impulse for the continued existence of the MPI, as speculated in recent literature,[99] is, however, rather unlikely. There was without a doubt a certain obligation towards Hahn and, certainly, some of the players in the MPG took particularly seriously the fact that the MPIC bore "the name of the honorary president"[100] when considering a possible closure.[101] The result, however, was not to preserve the institute at all costs for this reason, but instead the key was the proposal to disconnect from the name as long as possible before an eventual final stroke was delivered and then transfer it to another institute. Because of the focus on radiophysical and radio-

98 Lüst to Köster, April 3, 1967, in: AMPG, II. Abt., Rep. 62, No. 494.

99 See Andreae, Biogeochemische Forschung, 145.

100 Protocol of the 50th MPG Senate Meeting, March 12, 1965 in Berlin, in: AMPG, II. Abt., Rep. 60, No. 50.SP.

101 See Köster to Butenandt, April 14, 1967, in: AMPG, II. Abt., Rep. 66, No. 843.— See also: Andreae, Biogeochemische Forschung, 145.

chemical research, Carl Wagner suggested Karlsruhe for this move, since it was the location of the fast breeder reactor.[102]

In the end, there were good and rational reasons not to phase out the institute. Firstly, the contracts of the acting department directors, namely Hintenberger and Wäffler, ought to be allowed to expire. Although, as we have seen, Heisenberg considered waiting for exactly this, it would require general acceptance for a "phasing-out" period lasting more than ten years, since the contracts were not scheduled to terminate until the end of the 1970s. Affiliation of the department managers to other MPIs proved to be difficult, in particular in the case of Hermann Wäffler, since the working environment at the MPIC in Mainz was ideal for him; the linear accelerator that was a key component for Wäffler's research was right next door. The only alternative would have been to set Wäffler up at Mainz University but he was obviously not sufficiently integrated in the local environment, at least to the extent that the University would be willing to take this step.[103] Eventually, the option was discussed to make further appointments to the MPIC, but to limit the contracts to ensure the "motivation" of the new members. In addition to these considerations, it became increasingly clear that the MPG wanted to take future steps only with a specific orientation towards the research groups and fields of research of the neighboring university. The fact that Adolf Butenandt, among others, informed the dean of the university of Hartmann's rejection[104] was not the only evidence of this tendency. In particular, the active role of Hans Ehrenberg (1922 – 2004), who in the meantime

102 Wagner to Ziegler, March 18, 1964, in: AMPG, II. Abt., Rep. 62, No. 495.

103 See minutes of the "Future of the MPIC" committee meeting on March 5, 1966, pp. 4 f., in: AMPG, II. Abt., Rep. 66, No. 841; minutes of the "Future of the MPIC" committee meeting on February 24, 1966, pp. 3 f., in: AMPG, II. Abt., Rep. 66, No. 841.—See also: minutes of the "Future of the MPIC" committee meeting on April 6, 1967, pp. 4 f., in: AMPG, II. Abt., Rep. 62, No. 494.

104 See Butenandt to Riesler and the dean of the Univerity Mainz, January 20, 1966, in: AMPG, II. Abt., Rep. 66, No. 841.

was attending the "Future of the MPIC" committee meetings as the representative from Mainz University, was also a clear hint. Ehrenberg played a significant role in the establishment of the linear accelerator in Mainz and, together with Wolfgang Gentner, strongly promoted greater cooperation between the university and the MPI—successfully, as we will ultimately see. In the end it was Gentner who for the first time suggested an orientation towards the field of the ordinariate of meteorologist Christian Junge at the university in order to strengthen the cooperation between projects dealing with cosmochemistry at the MPI and the university in general. He felt that meteorites research itself had gradually passed its zenith and there was therefore a need to think in larger dimensions by strengthening a connection between aerospace and the international research programs linked to it.[105] Gentner initially proposed Johannes Geiss (born 1926) as a possible successor who could have assumed such a collaboration;[106] he was a student of Friedrich Houtermans, who, after several stays in the US, was once again working in Bern at the university of his former teacher.[107] During his time in Chicago, Geiss worked together with Friedrich Begemann (1927 – 2018), who in the meantime had become a close collaborator with Wänke at the MPI. With the appointment of Geiss, both Gentner and Ehrenberg hoped in particular for an advantageous connection with the

> "excellent meteorologist from the Mainz University, Mr. Junge … He [Junge] is not entirely satisfied regarding his opportunities there. He could work as 'freelancer' at the MPI and use the institute's facilities. Mr. Junge's research focuses on circulating processes in the higher atmosphere, an area where chemistry also

---

105 Minutes of the "Future of the MPIC" committee meeting on February 24, 1966, pp. 3 f., in: AMPG, II. Abt., Rep. 66, No. 841.

106 Ibid., 5 f.

107 See biographical information about: minutes of the "Future of the MPIC" committee meeting on March 5, 1966, pp. 6 f. in: AMPG, II. Abt., Rep. 66, No. 841.

> plays a significant role. All things considered, Mainz could develop into a center."[108]

Initially, this argument convinced the committee, and Ehrenberg and Gentner were asked to prepare documents about Geiss. They did so, but again things didn't go as planned and in the end the proposal that was initially considered very positive and had already been presented to the senate[109] had to be abandoned. As Werner Köster, the acting president of the "Future of the MPIC" committee reported, this result was, interestingly, mainly due to an objection from Christian Junge. A question from Otto Hahn, who also sat on the committee and wanted to know *"if this was to be seen as animosity towards the Max Planck Society"* was, however, answered in the negative by Gentner.[110] Hahn's inquiry regarding possible tensions between the MPIC and the university was not entirely without basis. In prior years, for example, there was no doubt that cooperation with the former second director of the MPIC, Fritz Straßmann, who was then at Mainz University *"would leave much to be desired for reasons that will not be discussed here in detail"*.[111] Employees at the university's linear accelerator also pursued the problematic developments in the appointments at the MPIC, not without a certain gloating.[112] In the case of Junge's objec-

---

108 Minutes of the "Future of the MPIC" committee meeting on February 24, 1966, pp. 6f., in: AMPG, II. Abt., Rep. 66, No. 841.

109 See protocol of the 55th MPG Senate Meeting on March 11, 1966 in Hannover, p. 22, in: AMPG, II. Abt., Rep. 60, No. 55.SP.

110 Minutes of the "Future of the MPIC" committee meeting on May 17, 1966, p. 2, in: AMPG, II. Abt., Rep. 66, No. 841.

111 Wagner to Ziegler, on March 8, 1963, p. 3, in: AMPG, II. Abt., Rep. 62, No. 495.—Together with Otto Hahn, Straßmann had discovered nuclear fission in 1938 at the Kaiser Wilhelm Institute for Chemistry in Dahlem (later the MPIC) and was later a substantial key player in the reconstruction of the MPIC in Tailfingen and Mainz.

112 See Horst Kant/Gregor Lax/Anja Heller: Interview with Günter Herrmann and Norbert Trautmann, April 28, 2012.

tions, however, good reasons played a decisive role. Exactly one week after Gentner had proposed Johannes Geiss, Fritz Houtermans died in Bern and Junge had expressed his suspicion to Gentner that Geiss very likely would become his successor—which also ultimately did happen. Nobody wanted to risk another defeat. Apart from this argument, Junge, as chairman of the appointment committee of Mainz University that was also contracting, had apparently clarified that *"he was not willing to give his consent to a coupling of the vacant chair with the Max Planck Institute through a combined role."* In principle, however, he would be willing to cooperate with the MPIC.[113] It should be noted in advance that while Junge didn't make a combined role involving a connection between the university and the MPI a condition, he did at least insist on remaining a faculty member in addition to his function as director of the MPIC. Another factor against Geiss was that he was Wänke's age and worked in areas very similar to those of the Meteorites Research Department and Otto Hahn had expressed his fear that an appointment of Geiss could *"hit Wänke hard"*.[114] Finally, he asserted his position that one could not dish up Geiss *"as director to the leading men at the institute"*.[115]

Recently, it has been assumed in the literature that interest in Junge only developed after discussions about the appointment of Geiss.[116] This, however, is not the case, since the quote above from the "Future of the MPIC" committee makes it clear that they came up with Geiss not least because his contact to Junge himself should be followed up on. However, Junge had explicitly emphasized his willingness to coop-

113 Minutes of the "Future of the Max Planck Institute of Chemistry (Otto Hahn Institute)/Mainz" committee meeting on May 17, 1966 in Munich, p. 1, in: AMPG, II. Abt., Rep. 66, No. 841

114 Meeting of the "Future of the MPIC" committee on June 20, 1966, in: AMPG, II. Abt., Rep. 62, No. 494.

115 Minutes of the "Future of the MPIC" committee meeting on April, 6, 1967, in: AMPG, II. Abt., Rep. 62, No. 494.

116 See Jaenicke, Erfindung, 197.

erate with the MPG in the context of the discussions about Geiss, and thus, the idea was born to consider Junge himself for the directorship at the MPIC. This time, however, the process was very cautiously advanced to avoid another failed attempt. Gentner spoke first to Christian Junge and then informed the other committee members by the end of February 1968 of the fact that he had accepted *"as far as possible to take over the MPIC, but under the condition that his professorship at the Mainz university would be filled by another meteorologist."* Mainz University reacted positively to this arrangement, especially as Junge would remain a member of the Natural Sciences Faculty.[117] Considering the appointment periods that are possible in the MPG, this time the process was brought to a successful conclusion in almost a ludicrously short time. Less than two weeks later, on March 5, 1968, the MPG Senate decided unanimously to allow the application of the president and appoint Christian Junge as director and scientific member of the MPIC; he was the first meteorologist at the MPIC.[118]

1.4 Restructuring of the MPIC: Laying the foundation stone of an Earth System Science Institute

Junge took the first steps towards a substantial restructuring of the institute immediately after being inaugurated, laying the foundation for decades of development during which the institution would gradually be developed into an Earth System Science Institute. The

117 Minutes of the "Future of the MPIC" committee on February 23, 1968, pp. 2 f., in: AMPG, II. Abt., Rep. 66, No. 843.

118 Minutes of the 59th meeting of the MPG Senate in Stuttgart, March 5, 1968, in: AMPG, II. Abt., Rep. 60, No 59.SP.—see in particular also materials for the appointment of Junge provided for the meeting, in: AMPG, II. Abt., Rep. 66, No. 841

central changes, already in place at the end of the 1960s, included the establishment of two new independent departments at the institute (Atmospheric Chemistry and Cosmochemistry) and, on a management level, the introduction of a board of directors to replace the traditional institute's director as the sole decision-maker.

Contrary to what has been stated in recent research literature, the Cosmochemistry Department had not yet been established in 1967 which was one year before Junge's inauguration.[119] At that time, the Meteorite Research Group, which later became the Cosmochemistry Department, was housed in Hintenberger's Mass Spectroscopy Department. Immediately after the successful appointment of Junge, the board of trustees saw a need to clarify the structure planning for the institute and obviously it had not clearly defined in advance whether in addition to the department of the new director, the "Meteorite Research Department" *"should also be an independent department in the future"*.[120] It was due to Junge's active efforts that Cosmochemistry became anchored as an independent section at the institute. As early as December 1968, he drafted a proposal for the establishment of the two new departments.[121] This was accompanied by an additional proposal for the appointment of Heinrich Wänke as director of the Cosmochemistry Department.[122] At the beginning of the decade, Wänke had been deemed too young for the position of *the* institute's director, but in the meantime he had obviously attained the "right" age for a position as director *at* the institute. At the end of the 1960s, the chances for an

119 Palme specifies April 1, 1967, for the establishment of the department and the inauguration of its head Heinrich Wänke. See Palme, Heinrich Wänke, 208.

120 Minutes of the MPIC Board of Trustees meeting on May 24, 1968, pp. 2 f., in: AMPG, II. Abt., Rep. 66, No. 849.

121 See proposal for the establishment of two independent departments at the institute in Mainz, December 16, 1968, in: AMPG, III. Abt., ZA 95, folder 5.

122 See proposal for the appointment of Prof. Dr. Heinrich Wänke (scientific member of the institute) as director of an independent department according to § 28 (para. 2 of the statute), December 16, 1968, in: AMPG, III. Abt., Rep. ZA 95, folder 5.

Fig. 2: Christian Junge

Fig. 3: Heinrich Wänke with rock sample

independent Department for Cosmochemistry were generally much better than a decade earlier, since now the faculty was headed towards a focus that was closely linked to one with increasing political and public interest, namely space research. NASA's Apollo Project represented an essential impetus. Research applications for projects that would be carried out with the hoped for moon rocks needed to be submitted well before the successful flight of Apollo 11. And, the "Meteorite Research Group", as well as Heinrich Hintenberger, had submitted a number of applications in advance. By March 1967, NASA had already promised the consignment of "moon material" as soon as the key objective of the Apollo program was achieved.[123] Seven of a total of eleven applications submitted from Mainz were approved. This was a good 1/17 of the 122 applications that were accepted by NASA overall in this connection.[124]

In response to Junge's proposals, the establishment of the two new departments was waved through with one abstention in the CPT section.[125] By March 1969, the MPG Senate had already negotiated this and also Wänke's appointment as director at the institute, as well as a further request from Junge aimed at restructuring the institute management.[126] In addition to these two departments, the independent sections of Heinrich Hintenberger (Mass Spectroscopy and Isotope Cosmology) and Hermann Wäffler (Nuclear Physics) continued to exist for Junge's entire term of office (1968 – 1978).

Soon after the restructuring at the level of the specializations, fundamental changes in institute management were introduced. While

123 Butenandt to Hintenberger, April 21, 1967 in response of a request by Hintenberger on March 21, 1967, in: AMPG, II. Abt., Rep. 66, No. 842.

124 See Hintenberger to Köster, March 31, 1967, in: AMPG, II. Abt., Rep. 66, No. 842.

125 See excerpts of the minutes of the Chemistry, Physics and Technology Section meeting, February 20, 1969, in: AMPG, II. Abt., Rep. 66, No. 844.

126 See protocol of the 62nd meeting of the MPG Senate on March 7, 1969 in Frankfurt am Main, in: AMPG, II. Abt., Rep. 60, No. 62.SP—the corresponding materials for the meeting can be found here: AMPG, II. Abt., Rep. 66, No. 844.

Junge was still facing negotiations on his later directorate with the MPG head office in Munich, interestingly, the board of trustees had no doubt that he would become *"director of the institute as a whole in any case"*.[127] Thus, the meteorologist would have been inducted in the tradition of the hitherto classical management style at the MPG that was essentially characterized by a single director as decision-maker of an institute. At the MPIC, this was also the case under Junge's predecessor, Josef Mattauch. Immediately after his inauguration, however, Junge dismantled these structures with the introduction of a board of directors, the members of which would each receive decision-making powers. This raised the question of why a newly appointed director at a MPI would, on his own initiative, limit his influence right at the beginning of his term. The first observation that arises from this statement is that the idea of a board of directors was not new with Junge at the MPG. After the failed attempt to appoint Hermann Hartmann as director in the mid-1960s, there had already been serious musings about the possibility of distributing management tasks. The real reason for this was specifically that Hartmann should be relieved of administrative tasks in his office.[128] For Junge, the board of trustees as well as the CPT section of the MPG pointed out in advance that he *"should decide the structure of the institute on his own ..."*;[129] nevertheless, the introduction of such a board of directors was already seen as a sure thing.[130] In addition to the administrative relief, scientific members already at the institute also apparently played a role, as they did not want to have their influence excessively restricted as a result of

127 Minutes of the MPIC Board of Trustees meeting on May 24, 1968, p. 2, in: AMPG, II. Abt., Rep. 66, No. 849.

128 Köster to Hartmann, April 5, 1965, in: AMPG, II. Abt., Rep. 66, No. 841.

129 Minutes of the meeting of the Chemistry, Physics and Technology Section on February 23, 1968, in: AMPG, II. Abt., Rep. 66, No. 843.

130 See minutes of the meeting of the Chemistry, Physics and Technology Section on February 23, 1968, in: AMPG, II. Abt., Rep. 66, No. 843.

the new appointment. Thus, they expressed concern to the "Future of the MPIC" committee that the institute's structure should not be settled until the inauguration of the new director. The *"scientific members of the institute ... felt that instead a board of directors should already have been mentioned during negotiations with Junge"*.[131]

In addition to these developments at the MPIC itself, it was apparent in the second half of the 1960s that reforms at a structural level had been considered in the MPG. This was a result of increasing public criticism of obsolete structures in the sciences that were seen as patriarchal[132] and in particular in the context of the so-called "Movement of 1968", which involved pointed criticism throughout all sectors of society of the academic structures in the FRG. Thus, at the end of the 1960s, the MPG addressed the question as to how it was, or should, be concerned by the university reforms.[133] The conversion of the Max Planck Institutes from the old, patriarchal management style to a board of directors must also be seen in this context. Although serious steps towards reforms in the Society were not actively advanced until 1969 and thus after Junge's appointment, it is likely that the introduction of a board of directors at the MPIC was seen as a welcome opportunity by the MPG in general to demonstrate the will to reform. On May 6, 1969, MPG president Adolf Butenandt and acting DFG president Julius Speer stated in a joint press release that the change of single institute structures would be the first step in ultimately reconsidering the organization of the MPG as a whole.[134] In fact, the MPIC was not the first Max Planck Institute to make a corresponding request to

131 Minutes of the "Future of the MPIC" committee meeting on February 23, 1968, in: AMPG, II. Abt., Rep. 66, No. 843.

132 See report of the president beffore the MPG Senate, March 7, 1969 in Frankfurt am Main, in: AMPG, II. Abt., Rep. 60, No. 62.SP.

133 Ibid., 9.

134 See minutes of the MPG Scientific Council meeting on June 13, 1969, in: AMPG, III. Abt., ZA 95, folder 5.

modify the management level. As early as June 1968, Lüst and Heisenberg had made a request to create a board of directors at the management level at the Institute of Extraterrestrial Physics.[135]

In the case of the Mainz MPIC, the institute's statutes, reissued in 1969, were also to be formed under the spirit of collegiality and were to be written specifically by *"Junge and the scientific members of the institute acting together"*.[136] The statutes came into force on March 7, 1969, and determined the purpose of the institute as follows:

> "§ 1: The institute is dedicated to basic research in the fields geo- and cosmochemistry, isotope chemistry and mass spectroscopy as well as nuclear physics".[137]

1.5 Summary of the Establishment of Atmospheric Chemistry at the MPIC

The appointment of a meteorologist and the establishment of atmospheric chemistry at the MPIC would have still been unthinkable at the beginning of the 1960s. The fields of specialization of the scientists that had been appointed up to that point together with a long list of other candidates who had been mentioned during preliminary negotiations, but whose names cannot be cited here are a testament to this fact. The sudden urgency for this at the conclusion of Junge's appointment is relatively easy to understand considering the many years of challenges. It is also significant that the atmospheric sciences enjoyed a far greater reputation at the end of the 1960s than at the beginning of the decade, and that the time was thus more favorable for the estab-

135 See Lüst and Heisenberg to Butenandt, June 5, 1968, in: AMPG, III. Abt., ZA 95, folder 5.

136 Protocol of the 59th meeting of the MPG Senate on May 5, 1968 in Stuttgart, in: AMPG, II. Abt., Rep. 60, No. 59.SP.

137 MPIC Statutes, March 7, 1969, here from: AMPG, II. Abt., Rep. 66, No. 850.

lishment of a department for atmospheric chemistry. Through the 1960s, considerable interest in environmental issues had gradually developed in both the sciences as well as in the political realm. In the US, major atmospheric science projects had already been initiated by the end of the 1950s and these also had an impact on Germany, though with some delay. Although at the beginning of the 1960s there was no significant engagement by the federal government in this respect, by at least around 1970, with the introduction in the FRG of the "Emergency Program for Environmental Protection", virtually every federal ministry was dealing with environmental issues.[138]

In the case of the MPG and the MPIC itself, a further crucial factor was that the new director represented a good compromise for integrating a young field of research that at the same time could be integrated with the structures anchored at the institute; this was particularly true for the mass spectroscopy methods.[139] Nevertheless, it took more than eight years to find an appropriate solution and the dilemma between tradition and innovation would not have been solved in the same way if one of the first appointment offers had been successful; adherence to the existing structures would have clearly taken precedence. The initial strategies for continuity at the institute were characterized mainly by the desire to pursue the main research areas practiced at that time, and also considerably, especially at the beginning, by the influence of retiring director Josef Mattauch. The development of a new research field played a subordinate role but, also in this framework, the notion of the "Harnack principle" of focusing on the appointment of a single scientist who was perceived as particularly suitable did persist. Especially in the case of Suess, the key criterion that an institute should

138 See Küppers et al., Umweltforschung, 127 – 155.

139 For the importance and development of mass spectrometry methods at the MPIC and their role at the research organizational level, see: Reinhardt, Massenspektroskopie. — See also continuing for the time from the late 1970s: Jochum, Drei Jahrzehnte.

be built around the needs of the director was bound to fail. Although Suess was provided with considerable funds, in 1961 he would have been subordinated to Mattauch for two more years. Thus, he not only would have continued the research focus of the former director, but to an extent he would also have been adopted into Mattauch's institute structures until Mattauch's retirement two years later. The latter would not hold true for Rudstam, although he also should have, and could have, made direct connections with the work in Mattauch's department.

The first failed attempts had made it painfully clear to the decision-makers that unconditional continuation of the existing institute structures in pure form was not possible. In the mid-1960s, the successor question, initially faced as an internal institute matter, was increasingly shaped by players not directly linked to the MPIC. More and more, high-ranking representatives from other MPIs who in the context of the CPT Section were jointly responsible for appointments at the respective institutes exerted influence, and it was only now that the establishment of new fields gradually came under consideration. Thus, it was initially agreed to establish theoretical chemistry as a new division previously not incorporated at the institute. When this attempt failed and there was increasing criticism from external entities, some of the players involved began to fear for the reputation of the Society as a whole and thus also for their own institutes—above all Werner Heisenberg. At the peak of the crisis even at the MPIC in Mainz new avenues were explored, including the core idea to orient more towards the neighboring university in Mainz.

A moment of prestige for the institute and the MPG as a whole that played an important role in the selection of some candidates—especially in case of Mößbauer—was to recruit top scientists from outside of the FRG, particularly if they had originally left the country for other places. Likely due to the urgency to act, any kind of "retrieval" motivation did not, however, play a role in Junge's appointment by that point. Instead, by the second half of the 1960s, the immediate proximity of the institute to Mainz University was a greater factor and, considering that closure of the institute had been seriously discussed, the "Future of the MPIC" committee gladly took advice from scientists based there. Gentner's idea to establish a center for cosmochemistry in Mainz, and the association with Christian Junge, were directly related to an orientation to the situation of the institute itself that was a result of these surroundings.

With his main focus on atmospheric chemistry, Junge satisfied the criterion (one that was in truth not always maintained) for a novel field of specialization. At the same time, his research corresponded to the methodological approaches practiced in the MPIC and was compatible with the idea of expanding the emerging field of cosmochemistry in Mainz. The general tenor was that with Junge, the new would be unified with what was already established. Indeed, it was acknowledged to the senate in the run-up to a meeting that took place in early March 1968 that

> "initially it [seems] surprising to propose a meteorologist as director of the Max Planck Institute for Chemistry. With Junge, however, the institute would find a head who developed a new field of research in meteorology [...] Atmospheric Chemistry uses research methods that are extraordinarily close to the Max Planck Institute for Chemistry."[140]

It was also emphasized to the MPIC board of trustees that with Junge *"a novel research direction [will be] introduced at the institute, which in some respects is closely linked to research that had long been promoted at the institute"*.[141] These links were based primarily on the methodological practices at the MPIC.

---

140 Materials for the senate meeting on March 5, 1968 in Stuttgart, p. 4, in: AMPG, II. Abt., Rep. 66, No. 841.

141 Minutes of the MPIC Board of Trustees on May 24, 1968, p. 3, in: AMPG, II. Abt., Rep. 66, No. 849.

The ultimate decision to go with atmospheric chemistry was probably largely due to chance, since the configurations offered locally in Mainz were suitable. The entire world had been searched for an ideal candidate and finally one was found right at the doorstep; he turned out to be the perfect link between preserving existing institute structures and establishing a specialist field that had never before been seated at the MPIC.

2

THE MPIC UNDER THE LEADERSHIP OF CHRISTIAN JUNGE, 1968 – 1978

As a consequence of Junge's appointment, a significant restructuring of the institution as a whole took place. This resulted in a strengthening of meteorite research, which had been poorly treated in previous years, both in the form of the new Cosmochemistry Department and in the opening of the first major research department in the FRG that specifically addressed the chemical composition of the Earth's atmosphere. The latter significantly influenced the development and expansion process of atmospheric sciences that gained momentum in the 1970s. But, before continuing with the further development of the MPIC, a brief biography of Junge is necessary.

Born on July 2, 1912, in Elmshorn (Schleswig-Holstein), Christian Friedrich Erich Junge showed an interest in chemistry at an early stage but then, for professional certainty, decided to study geophysics and meteorology.[142] Following his studies at the universities in Graz and Hamburg, he came to Frankfurt am Main, where he did his doc-

142 See Jaenicke, Erfindung, 188.—Junge's daughter confirms these reasons for Junge's decision: Gregor Lax: Interview with Heike Tilzer, on August 14, 2015.

torate under Franz Linke in 1935 and qualified as a university lecturer at the Meteorological and Geophysical Institute in 1952. Based on the available information, the political orientation of the young meteorologist within the context of the political turmoil at the beginning of the 1930s can only be assessed with great caution. A preliminary conclusion is that Junge presumably held a German nationalist position but did not sympathize with Nazi regime. He criticized the consequences of the Treaty of Versailles that were primarily derived from the reparation clauses included therein[143] and, in 1933 at the age of twenty, he joined the "Sturmabteilung" (SA) in Frankfurt am Main, where he was active in the "Sturmbann IV/63" and some other groups. He apparently never became an NSDAP (National Socialist German Workers Party) member and left the SA in 1935 to start work with the Reich Weather Service as an assistant to Franz Linke (1878 – 1944) in Frankfurt am Main.[144] At this time, he also got to know Helmut Landsberg (1906 – 1985), becoming his friend and establishing a good contact. Landsberg had done his doctorate under Beno Gutenberg (1889 – 1960) in 1930 and then, like Junge in later years, worked as an assistant to Franz Linke in Frankfurt. In 1934 Landsberg left Germany presumably as a refugee.[145] With the support of his former teacher, Gutenberg, who had already gone to the US in 1930, Lands-

143 Considering that the German chemical industry was at that time quite powerful, the decision initially seems surprising but is apparently true. Jaenicke refers to a direct statement made by Junge (see Jaenicke, Erfindung, 188) Junge's daughter also confirms this point (Gregor Lax: Interview with Heike Tilze, on August 14, 2015 in Konstanz) and Hans-Walter Georgii, meteorologist and Junge's first student, indicated that, soon after, he decided against studying chemistry for similar reasons (Gregor Lax: Interview with Hans-Walter Georgii, on April 27, 2015).

144 This resulted from an inquiry to the Federal Archive.—Email, Blumberg to Lax, August 15, 2014 and document with evidence, Dumschat to Lax, August 27, 2014.

145 References to this are provided by Jaenicke and Malone, with nothing explicitly stated, however. (Jaenicke, Erfindung, 191; See also the essay: Malone, Helmut E. Landsberg.) Landsberg himself later claimed to have merely wanted to learn English and to have emigrated to the U.S. for this reason alone (cf. Taba, Bulletin, 99).

berg received a professorship at Pennsylvania State College and also immigrated to the United States in 1934.[146]

Junge, however, remained in Germany and in 1942 was proposed within air force circles for a position as a senior civil servant for the war period; possibly due to his former SA membership, he was classified as ideologically reliable.[147] As a meteorologist, some of the places he was active included the war zones in North Africa, where he explored the weather conditions for missions of the German Air Force.[148] In France, he was involved in propaganda campaigns in which flyers were thrown from meteorological balloons.[149] Due to his activities during the Nazi era, he was interned by the Allies in 1945 but he was released as early as 1946.[150] After working in several institutions between 1947 and 1953—partly through mediation of contacts he had previously established in Frankfurt[151]—and re-qualifying as a university lecturer in 1952 at Frankfurt University, his most recent posting in Germany, Junge followed a recommendation from Helmut Landsberg and moved to the Air Force Cambridge Research Center (AFCRC) in Bedford, MA.[152] Later, he assumed US citizenship and stayed at the AFCRC until 1961; there, he dealt with the characteristics and distributions of aerosols in the Earth's atmosphere, which ultimately resulted in the discovery of the aerosol layer in the stratosphere that measures close to 10 km in thickness. This layer is primarily made up of sulfuric acid droplets[153]

146 See Ruth Prelowski Liebowitz (2008): Landsberg, Helmut Erich, in: Complete Dictionary of Scientific Biography, p. 196 – 200, URL: <http://www.encyclopedia.com/doc/1G2-2830905841.html>, status: May 23, 2018.

147 Document with evidence, Dumschat (Federal Archive Koblenz) to Lax, August 27, 2014.

148 Gregor Lax: Interview with Heike Tilzer, August 14, 2015 in Konstanz.

149 See Jaenicke, Erfindung, 188.

150 See Kant et al., Wissenschaftliche Mitglieder, 332 f.

151 Gregor Lax: Interview with Hans-Walter Georgii, April 27, 2015.

152 See Jaenicke, Erfindung, 191.

153 Junge and his colleagues published two papers on this discovery in 1961. In the first, they called attention to the aerosol cloud that was present in the stratosphere

and, in characteristic style of atmospheric research, is still today referred to as the "Junge-Schicht".[154]

Junge possibly had further options in the United States, but he decided to follow a call to the Institute for Meteorology at the Johannes Gutenberg University in Mainz and returned to the FRG in 1962.[155]

As a result of his formative influence on the establishment of atmospheric chemistry as a scientific field of investigation, Junge was already referred to as the *"discoverer of air chemistry"* by his former student Ruprecht Jaenicke.[156] Although Junge did play a key role in the establishment of integrative atmospheric research, this attribution seems inappropriate in several respects. First, it disregards other scientists who played an important role in establishing and developing atmospheric research in general. The notable names are legion, and include, among many others, Sydney Chapman, who shaped the classical theory of the distribution of ozone (discovered in 1839) in the atmosphere,[157] Bert Bolin at the MISU in Stockholm,[158] Charles David

around the world and later, in a second article, they furnished proof on the composition of this layer. See Junge et al., Stratospheric aerosols.—Junge/Manson, Stratospheric Aerosol Studies.

154 See Thomas Fleck: Wasserstoff-Emissionen und ihre Auswirkungen auf den arktischen Ozonverlust. Risikoanalyse einer globalen Wasserstoffwirtschaft, Jülich 2009, p. 81.

155 An indication of this arose within an interview with Junge's daughter; potentially complementary sources from the US could not be seen in the framework of this project. (Gregor Lax: Interview with Heike Tilzer, August 14, 2015.)

156 Jaenicke, Erfindung.

157 Christian Friedrich Schönbein discovered the ozone in 1839 as novel gas in the electrolysis of diluted acids (sulfuric acid and nitric acid) without knowing its chemical composition. This was detected by de Marignac and de la Rive, but not until some years later, see Claus Priesner (2007): Schönbein, Christian Friedrich, in: NDB Vol. 23, p. 384 – 386, URL: <http://www.deutsche-biographie.de/sfz78953.html>, status: May 23, 2018.—the molecular formula of ozone was finally formulated by Soret in 1865 and approved in 1867, see Rubin, History, 41.

158 Chapman, On ozone.—In 1929, a conference had taken place in Paris on the topic of ozone, where Chapman had presented his theory on stratospheric ozone. An elaborated article was published in the same year (Chapman, Theory).

Keeling, whose famous curve generated in 1957 on the linear CO_2 increase is known as the "Keeling curve" and John Hampson who added important components to Chapman's classical theory, to name but a few. Nevertheless, Junge played a major role, particularly in the context of atmospheric research that was formed and expanded in the FRG during the 1970s. When at the MPIC, he played an important role in the founding of the MPI for Meteorology in Hamburg that opened in 1975.[159] For the MPIC, however, Junge's appointment meant that the foundation had been laid for decades of development in which the institution was gradually expanded to form an institute for Earth System research.

In addition, as early as 1966, Junge together with Kurt Bullrich (1920 – 2010) from Mainz University and his former student Hans-Walter Georgii (born 1924), who had been a professor at the Institute for Meteorology and Geophysics at Frankfurt (IMG) since 1966,[160] had established a working focus for the study of atmospheric trace substances, from which the later DFG Collaborative Research Program 73 "Atmospheric Trace Gases" (SFB 73) emerged. This SFB was the first major DFG program on the composition of the Earth's atmosphere and was in place until 1985. Junge's department, lastingly, shaped the SFB 73 both in terms of the content of the research and its organization. Within its framework, institutional and cooperative structures were created at the beginning of the 1970s that initially were not equal, but were absolutely necessary for comprehensive atmospheric research.

159 Gregor Lax: Interview with Hans-Walter Georgii, April 27, 2015—Gregor Lax: Interview with Heike Tilzer, August 14, 2015. Junge compiled, among other things, a related memorandum for the CPT section of the MPG regarding the situation of meteorology and significantly supported the founding of an institute for meteorology and expressly called for the appointment of Hans Hinzpeter with reference to his own contacts with him. Cf. CPT section minutes, February 15, 1974, Bl. 6f., in: AMPG, II. Abt., Rep. 62, No. 1771, and June 17, 1975, Bl. 11, in: AMPG, II.Abt., Rep. 62, No. 1775.

160 Since 2005 Institute for Atmosphere and Environment (IAU).

In addition, instruments and methods were developed for the study of the chemical and physical nature of the Earth's atmosphere and novel perspectives on the atmosphere itself; these tools significantly shaped our current picture of the atmosphere. Together with the SFB, Junge's department advanced to become a basic structure for the future training of a new generation of scientists in the FRG.

Below, we will consider in more detail the beginnings of the new departments at the MPIC following Junge's appointment.

2.1 The beginnings of the departments for Atmospheric Chemistry and Cosmochemistry

Both new departments, Atmospheric Chemistry and Cosmochemistry, started with significant success in 1968. It would be wrong not to mention cosmochemistry at this point, as it was an important branch development, which remained firmly integrated in the institute until 2005. However due to the focus of this paper, it can only be outlined in brief here, but it would unquestionably serve well as a subject of its own right.

Based on the successes of the MPIC "Meteorite Research" working group mentioned above in participating in the Apollo program of NASA, it already can be suggested that, after Junge's appointment and the restructuring at the institute, further planning steps were in part strongly influenced by the needs of moon research that was on the rise at that time. After the successful moon landing of Apollo 11 in July 1969, the vessels containing moon rocks were subjected to an elaborate two-month quarantine process[161] before the samples could finally be delivered to the respective research institutes. Interestingly, it was not an employee of Wänke's group but rather Hans Voshage

161 See Helmut Hornung, Mond, 96.

from Heinrich Hintenberger's department who was delegated to fly to Houston[162] to receive the moon rock and return to the Mainz institute within 48 hours on September 18, 1969.[163] This haste was necessary, as the isotopes that were to be examined were subject to rapidly progressing degradation processes.[164]

The pressure to perform associated with the moon samples was high, since NASA continued to allocate samples only if a research group had shown successes with previous samples.[165] Even before the first sample reached the institute, however, Heinrich Wänke in Mainz had no doubt that his *"working group … [would be] busy with the moon material for several years, in particular with material from later flights"*.[166] Palme vividly describes that the obligation to succeed with the limited amount of material was productive in pushing the improvement of geochemical and isotopic as well as mineralogical working methods, with regard to both the measurement and analysis methods and the reduction of the amount of material samples required.[167] Continuous optimization of the measuring instruments was one factor that had positive effects in the long term, particularly at the MPIC, and resulted in a permanent extension of the instruments available there; this consequence also benefited the Department for Atmospheric Chemistry.[168] In 1970/1971, these positive effects included major improvements in the measurement of main and trace substances in meteorites and significant modifications to the mass spectrometer MS7, in which photographic plate detection

162 Carsten Reinhardt/Gregor Lax: Interview with Friedrich Begemann, January 6, 2012.

163 See Hornung, Mond, 96; Palme, Heinrich Wänke, 218.

164 See Palme, Heinrich Wänke, 218.

165 See ibid., 220 f.

166 100 Gramm Mond, 105.

167 See Palme, Heinrich Wänke, 220 f.

168 Carsten Reinhardt/Gregor Lax: Interview with Friedrich Begemann, January 6, 2012.

that was conventional at that time was enhanced with the option for electrical detection.[169]

The Mainz institute received the largest quantity of moon rock ever awarded by NASA to an institution outside of the US. The sample weighing 105.9 g that was brought to Mainz by Voshage in mid-September was soon followed by another sample of 140 g in October.[170] The successful collaboration between the Department for Cosmochemistry at the MPIC and NASA was the main reason that Wänke and his employees received a 50,000 DM budget increase for the year 1970 from the MPG and even Christian Junge agreed to hand over an additional 30,000 DM from his own budget for the work on the moon rock in the Department of Mass Spectroscopy.[171] Furthermore, in 1970, four of eleven new positions were assigned to the Department for Cosmochemistry. As well as another position for Heinrich Hintenberger's department, two of them were promised based on the work on the Apollo program alone. Initially, however, only one new position was targeted for Junge's department.[172]

This does not mean that the Department for Atmospheric Chemistry did not start successfully—on the contrary. The opening of the DFG Collaborative Research Program "Atmospheric Trace Gases" established by Junge, Hans-Walter Georgii and Kurt Bullrich in 1968 was central for the developments of the following years. At this point, Georgii and Bullrich must be briefly introduced.

Like Junge, Georgii had originally wanted to become a chemist but, due to the poor conditions for studying chemistry after the Second World War, he decided to register for physics and meteorology in

169 See Jochum, Drei Jahrzehnte, 4 f. and 11.

170 See Hornung, Mond, 96 f.

171 See note Roeske, November 18, 1969 to the board meeting imminent on November 20, 1969, p. 4, in: AMPG, II. Abt., Rep. 66, No. 850.

172 Ibid.

Frankfurt am Main.[173] There he became Junge's first graduate student; Junge advised via mail from the US later during his dissertation, which was officially supervised by Ratje Mügge (1896 – 1975).[174] Mügge was then Head of the IMG[175] in Frankfurt[176] and, at the beginning of the 1950s, probably also helped Junge after his release from captivity as a prisoner of war to get the position at the weather service in Hamburg.[177] In 1965, Georgii finally got a posting as a chair at Frankfurt University where he became head of the IMG.[178]

Kurt Bullrich had also studied physics and meteorology in Frankfurt, where he graduated in 1942. He then spent time as an assistant at the new Meteorological and Geophysical Institute of Mainz University in 1949, where he qualified as university lecturer in 1963 and was awarded a professorship in 1968 (the year in which SFB 73 was founded). His focus was atmospheric radiation and this also became his main research field at the SFB 73, where he addressed the direct influence of aerosols on the radiation budget of the atmosphere.[179] Collaborative research programs had just recently been established as a funding instrument in 1968, and the SFB 73 was the first in-depth research program in Germany to address a comprehensive investigation of the chemical composition of the Earth's atmosphere and its mutual relationships with the geo-, hydro-, and biosphere.

---

173 Gregor Lax: Interview with Hans-Walter Georgii, April 27, 2015.

174 Ibid.

175 Since 2005 Institute for Atmosphere and Environment (IAU).

176 See Hofmann, Mügge; URL: <http://www.deutsche-biographie.de/pnd133807681.html>, status: May 23, 2018.

177 Gregor Lax: Interview with Hans-Walter Georgii, April 27, 2015.

178 See Kürschners Deutscher Gelehrtenkalender. 22. Ausgabe 2009. Berlin: K. G. Saur 2011, p. 1153.

179 Ruprecht Jaenicke: Laudatio auf Kurt Bullrich, anlässlich seines Todes am 31.3.2010, URL: <https://www.blogs.uni-mainz.de/fb08-ipa/files/2014/07/Laudatio_bullrich.pdf>, status: May 23, 2018.

From the outset, Junge's department was one of the pillars of this ambitious project and the research activities carried out there were tightly interwoven with the objectives of the SFB 73. Therefore it is necessary to outline the approaches and structures of the collaborative research program at this point.[180]

Peter Warneck, an employee of Junge and later a founding director of the Leibniz Institute for Tropospheric Research in Leipzig, brought the views of cycles, sources and sinks among the Earth's spheres directly back to Christian Junge[181] and indeed a corresponding approach can already be found in his monograph of 1963, in which Junge also refers at times to other authors.[182] However, much earlier, similar considerations had already been made, some by prominent scientists, including as early as in the 1920s by Russian geologist and chemist Wladimir I. Vernadsky (1863 – 1945).[183] In contexts that were firmly based in atmospheric research, an article by Roger Revelle and Hans Suess from 1957 must be named; it not only addressed the CO_2 cycle between the atmosphere and the oceans but also emphasized the role of humanity as an increasingly important source of CO_2.[184] Also worthy of note is the definition provided by Stockholm meteorologist Bert Bolin, who in 1959 described "atmospheric chemistry", which at that point had still not been solidly established as a field, as an *"interplay between the atmosphere, the surface of the continents and oceans, biological activities, man and last but not least the ever moving atmosphere itself"*.[185] The concept of cycles as well as sources and sinks for substances relevant to the atmosphere became a central starting point for the SFB 73

180 As already indicated in the introduction, the following statements are mostly based on the published article: Lax, Aufbau.

181 See Warneck, Geschichte, 3.

182 Junge, Air Chemistry, 21 and 34.

183 See Andreae: Biogeochemische Forschung, 134f.

184 Revelle/Suess, Carbon Dioxide Exchange.

185 Bolin, Atmospheric Chemistry, 1663.

Fig. 4: Junge's Working Group 1972

research program and above all for Junge's department. This would be the cornerstone for the future orientation of the entire MPIC and its development up to the present. Appropriate departments were then established at the institute which addressed and still today address the geo- and biosphere with a continuous view to their exchange relationships and interrelations with the atmosphere.[186]

The SFB lasted fifteen years and reached the maximum period possible for funding of collaborative research centers. Immediately thereafter, in 1985, Georgii and others initiated the SFB 233 "Dynamics and Chemistry of Hydrometeors" which was funded beginning in 1986 and which specifically addressed the "wet" atmosphere (in particu-

186 A first step in this direction was carried out in 1987 with the establishment of the Biogeochemistry Department since then managed by Meinrat O. Andreae.

lar the ice phase, hail creation and fog spread and formation).[187] As early as 1966, what became known as the "iron triangle"[188] (the Universities of Frankfurt, Darmstadt and Mainz) had a common focus in the study of atmospheric trace substances[189] In 1968, the SFB 73 arose from this initiative. Because of Junge's move from Mainz University to the MPIC, the MPI was integrated from the beginning into the collaborative research program in addition to the other universities. Between 1970 and 1985, the involved institutions received funding that over the years amounted to 26,591,000 DM.[190] The structure of the SFB was not self-evident, because Max Planck Institutes had no priority in the DFG distribution of funds since—in contrast to the universities—they had (and still have) their own large research budgets. Partly because of the healthy financial position, the MPIC became a sought-after address even for employees from university institutions working together in the SFB.[191]

In any case, the department for Atmospheric Chemistry was to be kept on board, although because of the university-oriented focus of the DFG, Mainz University was appointed as the primary provider for the collaborative research program. The main spokesperson, who changed every two years, was always a member of one of the participating universities. Junge never served in this position and even in

187 See DFG (ed.): Jahresbericht der DFG, Bd. 2, Bad Godesberg 1986, pp. 839 f.—Georgii verified his participation at the SFB 233 (Gregor Lax: Interview with Hans-Walter Georgii, on April 27, 2015).

188 See comment by Helfer, February 4, 1970, in: BA B227/010802 (DFG-Sonderforschungsbereich 73).—The "iron triangle" mainly consisted of very good contacts at a personnel level. Georgii (Frankfurt) had been Junge's (Mainz) student; both had studied in Frankfurt and in turn there had been good contact between Frankfurt and Darmstadt in the past. Thus, Ratje Mügge, Georgii's doctoral supervisor (also Frankfurt) had taught in Darmstadt every now and then. (Wolf, Verzeichnis, 143).

189 See details on the center of gravity of atmospheric trace substances, February 17, 1966, in: BA B227/010802 (DFG-Sonderforschungsbereich 73).

190 DFG, Jahresbericht 1985, 883.

191 Gregor Lax: Interview with Hans-Walter Georgii, April 27, 2014.

1972, when he was explicitly proposed for it by Georgii at DFG, it was not accepted.[192] But as we shall see, the Department for Atmospheric Chemistry nevertheless shaped the research focuses of the DFG program for the long term. One of the most important starting points for the foundation of the SFB 73 was a concerted cooperation among several smaller institutions to establish a research structure that could keep in step with international competition that was perceived to be "overwhelming".[193] In fact, the development of atmospheric research in the FRG lagged behind in particular the US,[194] where as early as the second half of the 1950s, suitable projects had been initiated, some on a large scale. At this point, only the "National Atmospheric Program" initiated in 1959 and of course the foundation of NCAR that started work in Boulder (Colorado) in 1960/1961 will be named.[195]

In the FRG, however, there were still problems establishing any kind of effective cooperation with those research institutions that were identified as institutional partners for the atmospheric research program. This issue was clearly reflected in the criticism from the reviewers during the SFB establishment years (1968–1970), which identified only inadequate collaboration between the few institutions that were already participating. The reviewers perceived that project planning was carried out in parallel and not in coordination with the institutes involved.[196] Another criticism that arose several times at the beginning of the 1970s was the lack of a contingent of scientists explicitly trained

192 See Georgii to Pestel, December 19, 1972.—Kurt Bullrich became speaker instead. See note Wilken to Kirste, August 27, 1973. Both in: BA 227/011244 (Sonderforschungsbereich 73).

193 See also: Christian Junge, November 21, 1967, Remarks to the Collaborative Research Center "Atmospheric Trace Substances", in: BA B227/010802 (DFG-Sonderforschungsbereich 73).

194 Schützenmeister: Zwischen Problemorientierung, 109.

195 See Hart/Victor, Scientific Elites, 650.

196 Stellungnahme der Fachreferate zum SFB, January 5, 1970, in: BA B227/010802 (DFG-Sonderforschungsbereich 73).

as chemists within the SFB 73. The first staff listing submitted to the DFG contained only a single chemist and he was merely a doctoral candidate serving as a guest scientist in the FRG at the time.[197] In light of this criticism, Junge and Georgii were again obliged to revise their employee lists and, on this occasion in particular, to emphasize the presence of chemists in their working groups.[198] Jürgen Hahn, Rudolf Eichmann und Gregorios Ketseridis were now listed for the MPIC department.[199] Another problem identified by the experts was the lack of adequate networking with other institutions where the focus was a classical field of chemistry and not meteorology, as was largely the case for the working groups of the "iron triangle" at that time.[200] In particular, a working group from analytical chemistry was felt to be lacking by the DFG and Hans-Walter Georgii himself also expressly acknowledged this deficit.[201] Initially, there appeared to be no satisfactory solution.[202] According to Peter Warneck, this problem resulted from the extreme challenges for chemists in detecting trace substances in the air because initially they had only limited "tools" available to them.[203] Chemical as well as physical methods and instruments, for instance in the field of mass spectroscopy,[204] existed in principle but first needed to be adapted to the requirements of atmospheric research.[205]

197 Ibid.—This was probably Gregorios Ketseridis from Greece who worked in Ruprecht Jaenicke's group on organic components of aerosol particles in pure air.

198 Junge to Baitsch, January 29, 1970, in: BA 227/10803 (DFG-SFB 73).—January 30, 1970, Georgii to Baitsch, in: BA 227/10803 (DFG-SFB 73).

199 See ibid. Junge to Baitsch, January 29, 1970, in: BA 227/10803 (DFG-SFB 73).

200 Amongst others here: Streiter, statement on the SFB 73, Juli 14, 1970, in: BA 227/010919 (DFG-SFB 73).

201 See Georgii to Kirste, September 18, 1972, in: BA 227/011037 (Sonderforschungsbereich 73).

202 See opinion on the SFB 73, March 27, 1974, in: BA 227/011460 (Sonderforschungsbereich 73).

203 Warneck, Geschichte, 6.

204 See for the history of mass spectroscopy at the MPIC: Reinhardt, Massenspektroskopie.

205 Gregor Lax: Interview with Hans-Walter Georgii, April 27, 2015.

The initially problematic situation at the SFB was probably also a result of a rather weak impetus on the part of the institute for interdisciplinarity implemented on a practical level. Atmospheric research was apparently not recognized or perceived as a legitimate field of activity by many "classical chemists". Junge, at least, did make it clear to the DFG reviewers that chemistry institutions often showed no interest in the respective working areas.[206] Indeed, at least initially and even within the MPIC, some employees expressed certain skepticism about the appointment of a meteorologist to the institute.[207] However, this suspicion towards other disciplinary areas was based on reciprocity, as was revealed at a meeting for the further structuring of the SFB in 1970. Here, members expressly spoke out against the inclusion of chemistry institutes, since they felt other interests were the focus at these institutes and chemists would first require an introduction to meteorological problems.[208] It can therefore be concluded that the MPIC was mainly accepted as an SFB member because it was represented by a renowned meteorologist and not because it was felt that a dedicated institute for chemistry should be included from the outset. However, the SFB members could present the involvement of this kind of chemistry institution to the DFG. In the period that followed, the DFG experts really encouraged the institutions involved in the SFB to put more emphasis on integrating chemistry. Finally, despite the criticism mentioned above, the proposal was accepted for funding, but with the conditions first to involve more chemists and second to build a larger network of cooperation.[209] In the following years, active collaboration with other institutions was among the notable features of the SFB and the participating institutes. Although the distribution

206 See Stellungnahme zum SFB 73, July 14, 1970, Streiter, handwritten note, in: BA B227/010919 (Sonderforschungsbereich 73).

207 See C. Reinhardt/Gregor Lax: Interview with Friedrich Begemann, January 6, 2012.

208 See note Helfer, February 4, 1970, in: BA B227/010802 (Sonderforschungsbereich 73).

209 See ibid.

across disciplines was undeniably imbalanced at the beginning, on the whole over the years, the member pool increasingly touched on a wide range of subjects, from physics and chemistry of the atmosphere to geochemistry and photochemistry as well as theoretical meteorology and geology.

Despite the initial difficulties, the chemistry branch was to advance to become the important driving force for the research carried out in the SFB. Over a long period of time, the groups of the collaborative research program were clearly dominated by the Atmospheric Chemistry Department of the MPIC. Ten of the twenty employees specified in the research proposal for the years 1971 – 1973 came from there; the Universities of Frankfurt and Mainz were only represented by five employees.[210] As of 1974, the MPIC provided even more than half—twenty out of thirty-nine—of the employees; in contrast, the Universities of Frankfurt and Mainz each had nine and Darmstadt TH only one.[211] This remained the situation in the years that followed and the TH was apparently virtually inactive in the second half of the 1970s, although it remained a part of the SFB.[212] The real pillars of the SFB at the institutional level were an "iron triangle" (MPIC and the Universities of Frankfurt and Mainz) but not on a geographic level (Mainz, Frankfurt, Darmstadt).

The research program of the SFB was based on a widely diversified approach. According to the official self-description, the goal was to

> "cooperatively and comprehensively study the chemical composition of the atmosphere and its global and long-term change. The exploration of sources and sinks for individual compounds, as well as the interaction between the ocean and the atmosphere on

210 See the alphabetical Liste der Mitarbeiter im SFB-Antrag on June 19, 1970, in: BA B227/010919 (Sonderforschungsbereich 73).

211 See Liste der Mitarbeiter im Fortführungsantrag des SFB 73 for the years 1974 – 1976, chapter 1.4, in: BA B227/011461 (Sonderforschungsbereich 73).

212 See president of the Darmstadt TH to the DFG president on July 18, 1979, in: BA B227/077088 (Sonderforschungsbereich 73).

Fig. 5: Manfred Schidlowski's Working Group "Paleoatmosphere": Peter Appel, Jürgen Hahn, Manfred Schidlowski, Hans Wong, Rudolf Eichmann (from left)

> the one hand and the biosphere and the atmosphere on the other hand [are paramount, GL] … Particular significance is thus paid to research projects that address the chemical evolution of the atmosphere."[213]

The first item to note here is the explicit reference to the interaction between the three Earth spheres. In addition, there is unambiguous reference to the *chemical* composition of the atmosphere, while physical investigations of clouds are not explicitly mentioned. The interest in the history of the origins and development of the atmosphere stressed at the end of this description bore a direct relationship to Junge's department, where geochemist Manfred Schidlowski was investigating the palaeoatmosphere. Although this topic played, and still plays, a key role particularly in mapping climate changes for long time periods,

213 See DFG, Jahresbericht 1970, 599.

and the SFB stressed the importance of the palaeoatmosphere, Schidlowski's group was always a little bit isolated in the SFB due to their work that focused primarily on the geosphere.[214]

In 1970, the collaborative research program still had a structure that was somewhat confusing and consisted of a total of 11 individual project groups. A common aspect of all projects was that insight should be provided into global distributions, sources and sinks for the trace gases and aerosols that were of interest. Although environmental issues, particularly those with regional provenance, played an increasing role in the FRG and even in public discourse through the 1970s which even forced the German government to intensify its environmental policy,[215] small-scale phenomena, such as local air pollution, did not attract much attention in the SFB. Instead, global distribution of aerosols and trace gases was at the forefront.[216] The situation stayed that way throughout the decade-and-a-half lifetime of the SFB. Whereas Junge and Georgii had addressed topics with particular relevance for policy in addition to the SFB, including the issue of "acid rain" for instance, projects in the collaborative research program always remained focused on basic research into the chemical composition of the atmosphere and its interactions with the geo- and biosphere. A direct coupling of the research to environmental issues with immediate political relevance was probably less necessary because a relatively generous financial basis was created by the DFG foundation and thus dependency on (earmarked) third-party funds was low.[217] However, a slight shift arose in the middle of the decade when anthropogenic influence on the atmosphere as a global factor

214 See Fortführungsantrag für 1974 – 1976, chapter 1.4, in: BA B227/011461 (Sonderforschungsbereich 73).

215 Müller, Innenwelt, 73.

216 See Protokoll der Gutachtersitzung des SFB 73, November 7, 1972, in: BA 227/011244 (Sonderforschungsbereich 73).

217 Gregor Lax: Interview with Hans-Walter Georgii, April 27, 2015.

attracted further attention, even in the SFB, but this did not lead to a significant change in the objectives that had been formulated at the outset for the program either. Instead, for example, Junge argued in 1974 for a Medical Biology section at the MPG:

> In order to understand and assess the influence of human activities on the composition of the atmosphere, it is necessary to have a clearer understanding of the cycle of trace substances in the systems of the atmosphere, the ocean and the Earth's surface; specifically, there should be focus on an atmosphere unaffected by humans.[218]

In addition to the "Paleoatmosphere" Group, the MPIC was involved with three other groups whose topics would later make a marked impact on the SFB. These groups included the "Photochemical Reactions in the Atmosphere" group headed by Peter Warneck, Ruprecht Jaenicke's "Constitution of the Atmospheric Aerosol" group and the "Atmospheric Trace Gases (CO, N_2O, H_2)" group under the leadership of Wolfgang Seiler.[219] A point of interest about Seiler's group is that as early as 1970 a more comprehensive range of trace gas research was planned that later became an integral aspect of the SFB.[220] In addition to the gases already mentioned, sulfur dioxide (SO_2), hydrogen sulfide (H_2S), mercury (Hg), ammonia (NH_3), nitrogen compounds, hydrocarbons and of course ozone (O_3) were also included and were ultimately added to the official description of the SFB in the publicized DFG annual reports.[221] In the decades that followed, Warneck, Jaenicke and Seiler played important roles in the further development of the FRG's atmospheric scientific research. As already noted, Warneck in later years was heavily involved in the establishment of the

218 Junge to Aschoff, January 4, 1974, in: AMPG, III. Abt., ZA 95, folder 3.

219 DFG, Jahresbericht 1970, 599.

220 DFG-Antrag 1970/71, October 1, 1969, by Christian Junge, p. 17, in: BA B227/10803 (DFG-Sonderforschungsbereich 73).

221 DFG, Jahresbericht 1976, 779.

Leibniz Institute for Tropospheric Research in Leipzig and became its founding director when the institute opened in 1992.[222] Wolfgang Seiler's contributions included the establishment of standards for research on the global distribution of trace gases and he became director of the Fraunhofer Institute for Atmospheric Environmental Research in Garmisch-Partenkirchen in 1986.[223] Jaenicke eventually took a professorship at the Institute for Physics of the Atmosphere at the Mainz University in 1980, where he focused in particular on aerosol research, which had already been his major area during his work at the SFB 73.

The collaborative research program itself did not remain in its original structure for long, but was fundamentally redesigned in the context of a renewal application as early as 1973. The existing working groups were supplemented by others and these were no longer listed as individual projects. Instead, four comprehensive subject areas, each with one main leader, were established with the previous groups assigned as subproject groups. One reason for this change was that, in the course of the new accreditation, the DFG experts wanted more transparent conditions for the financial structure and this was an attempt to counter what was a far too flexible "shifting" of money between individual groups.[224] The four new comprehensive subject areas now precisely matched the research focuses of Junge's department and three of them were eventually headed by MPIC people. Junge himself was now responsible for the main area, "Trace Gases", which included the groups "Global CO_2 Balance", "Measurement of Trace Gases" and "Propagation of Trace Gases and Their Exchange Processes Between Water and Air". Peter Warneck headed the subarea "Physicochemical Processes in the Atmosphere", which in particular

222 See Leibniz Institute website, URL: <https://www.tropos.de/institut/ueber-uns/das-institut/>, status: May 23, 2018.

223 See Trischler/vom Bruch, Forschung, 150 f. and 153 f.

224 Gregor Lax: Interview with Hans-Walter Georgii, April 27, 2015.

worked on rainfall chemistry and reactions of gases and radicals in the atmosphere as well as methods to detect free radicals. The third subarea connected to the MPIC Department consisted of Schidlowski's "Paleoatmosphere" group. The only topic complex that was not headed by a direct employee of Junge was the "Aerosols" field; Kurt Bullrich had undertaken this task. However, Ruprecht Jaenicke from the Atmospheric Chemistry Department was integrated in this group; after starting his ordinariate at Mainz University in 1980, he had finally replaced Bullrich as head of the aerosols project in 1981.[225]

Despite more minor changes in the appointment of the subgroup leaders, the personnel structure on the overall management level and the distribution of the topics into four main areas remained unchanged until 1979 after the premature retirement of Junge, which was initiatd in 1978. At this time he took a backseat in the SFB as well, although he remained a member until his membership expired in 1985. Hans-Walter Georgii assumed the management of the "Trace Substances" group in 1980.[226]

The interdisciplinary approach of atmospheric research inevitably resulted in a cooperative structure, on not only a national but also an international level. The networking process connected with this collaborative approach will be illustrated in brief below using the example of the MPIC. Of course, there had been individual collaborations at the institute in the past; for example, the former Department for Mass Spectroscopy led by previous institute director Josef Mattauch had worked cooperatively with physicist Aaldert Wapstra at the "Instituut voor Kernfysisch Onderzoek" in Amsterdam (IKO). In the 1970s, however, the MPI started to network with other institutions on a much larger scale. Figure 6 shows lasting collaborations that were maintained by the Department for Atmospheric Research between 1968 and 1977 (just prior to Junge's early retirement). The graphic is based on a list

225 See DFG, Jahresbericht 1980, 790 and DFG, Jahresbericht 1982, 804.

226 DFG, Jahresbericht 1980, 790.

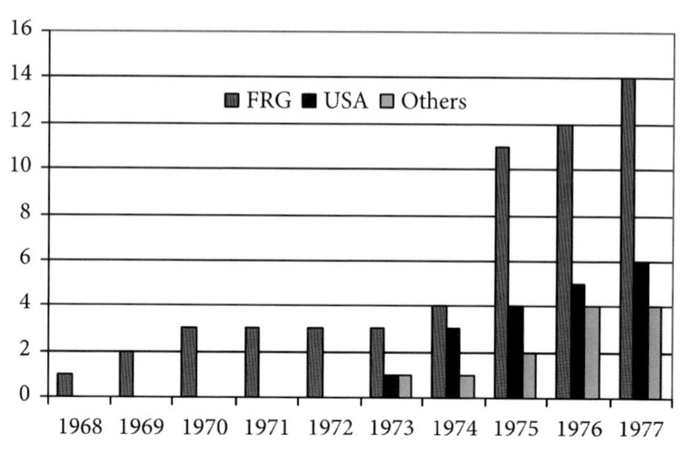

Fig. 6: Long-term collaborations of the Department for Atmospheric Chemistry, 1968 – 1977

from the department dated May 18, 1977,[227] naming those institutes and scientists with which there were intensive collaborations. Less continuous, but not necessarily unimportant collaborations, are not included. One example of note is NCAR in Boulder, Colorado with which very good relations had already existed since the beginning of the 1970s. Among other projects, MPIC employees had carried out joint research flights with their colleagues at NCAR in the framework of the SFB 73 in 1973.[228] Later on, there was also increasing exchange with Junge's successor as director of the Department for Atmospheric Chemistry at the MPIC, Paul Crutzen, who had worked at NCAR since 1974. Although it cannot be claimed that Fig. 6 is an exhaustive list, it reflects the impressive global networking by the Department for Atmospheric Chemistry throughout Junge's term which created the

227 List: Zusammenarbeit mit anderen Forschungsinstituten der Abteilung Luftchemie, May 18, 1977, in: AMPG, III. Abt., ZA 95, folder 4.

228 Georgii to Wilke, November 23, 1972, annex, in: BA 227/011460 (Sonderforschungsbereich 73).

infrastructure necessary for its contribution to the ambitious major project of researching the chemical composition of the atmosphere.

It is apparent that collaborations were predominantly developed with German (FRG) or American organizations, however occasionally cooperative research was undertaken with institutions in South Africa, England and France as summarized in the figure under "Others". In particular, there were links between the research on trace gas cycles by Wolfang Seiler's group and joint projects around the world. A real push for new collaborations arose in 1973, when the renewal application was granted and the four superordinate research priorities mentioned above were formulated.

2.2 The SFB 73 and the MPIC in the context of the consolidation of atmospheric sciences in the FRG

Starting in 1973 at the latest, the main focus of research at the SFB was significantly influenced by questions relating to chemistry (photochemistry, geochemistry-based paleoatmosphere, trace gases and aerosols). However, at an early stage and through the years, an increasingly heterogeneous composition of members in the working groups was evident.[229]

Although a detailed overview of the SFB research was provided earlier,[230] some of the key areas will be examined again below from specific points of view. Particular attention will be paid to those things mentioned at the start that led to the expansion and consolidation process of the atmospheric sciences that took place through the 1970s

229 See Liste der Mitarbeiter im Fortführungsantrag des SFB 73 for the years 1974–1976, chapter 1.4, in: BA B227/011461 (Sonderforschungsbereich 73).

230 Warneck, Geschichte.

(canonization, development of methods and instruments etc.). In addition, the connection and expansion potential of atmospheric chemistry research will become clear. As early as in the first half of the 1970s, this increasingly led to the emphasis on the "open big-science" approach that crossed disciplinary boundaries and lastingly shaped atmospheric research.[231] Furthermore, the early research interest in anthropogenic influences on substance cycles that at the time were still deemed "possible" was addressed. With all emphasis on the basic research orientation at the SFB, this subject was again compatible with questions that had already become relevant in public and political contexts in the first half of the 1970s. Only the discourse on pollution and the laying of a foundation for environmental policy in West Germany based on the "polluter pays" principle will be mentioned here.[232]

For the most part, the research carried out during the start-up period of the SFB 73 was shaped by two features. The first was that inventories of specific research areas were first created and the opportunities to relate them to and find relevance for them with other fields needed to be demonstrated. In particular, early on Junge had begun to generate basic atmospheric scientific research literature. In 1963, he had already authored the influential monograph on "Air Chemistry and Radioactivity", the key contribution of which was the preparation and first compilation of a wide range of literature from different

231 The term is based on the concept of "big science", differing however in its feature of having a decentralized structure and no centering of the costs in one location, in addition to high personnel and material expenditure. This structure is the result of necessity in atmospheric and also Earth-system research, as no individual institution is adequately equipped and specialized to process the entire range of research of the atmosphere/Earth System. Large programs and networks are thus a major feature of these areas. Schützenmeister describes the concept of open large-scale research taking atmospheric chemistry as an example. See: Schützenmeister, Offene Großforschung.

232 See Müller, Innenwelt, 74.

disciplines.[233] The book can be seen as a foundation stone for a canon of scientific literature relating to atmospheric chemistry.[234] It even garnered interest within the scientific community in the Soviet Union and in 1965 was translated into Russian.[235] At the end of the 1980s, the monograph was still cited as the only standard reference summarizing literature on atmospheric research.[236] The main focuses described within it included aerosol and trace gas research, which also became key topics of the SFB 73 later on. With reference to the material cycle of CO_2 Junge particularly stressed the importance of the interactions among the geo-, bio- and atmospheres.[237] He continued to retain the summarizing style of this monograph, in particular with a focus on aerosols research. For example an article with the same format and an overview-like character was written in 1968. In that article, Junge highlighted the importance of atmospheric chemistry for phenomena based on cloud physics, something he had repeatedly stressed since the early 1950s,[238] although such phenomena, as noted above, were never explicitly emphasized at the SFB. Work that thus referred to the current status of research and implicitly called for appropriate follow-up work continued into the subsequent period.[239]

Of particular interest is also the format with which new research results were presented in the early 1970s. In 1970 Junge published some research results together with Eugene McLaren which were based on an analysis of measurements of cloud condensation nuclei spectra and the chemical composition of 88 samples collected over a

233 Junge, Air Chemistry.

234 Also in: Jaenicke, Erfindung, 187; Andreae, Biogeochemische Forschung, 146.

235 Junge, Chimičeskij.

236 E.g. Gordon, Reviews, 1263.

237 Junge, Air Chemistry, 21, 23 and 29 f.

238 Junge, Survey, 592 ff. Jaenicke describes the story of Junge's equation for the determination of aerosol size distribution, which was attributed a status akin to natural law up until the early 1970s (Jaenicke, Erfindung, 190 f.).

239 See: Junge, Our Knowledge; Junge, Atmospheric Aerosols.

period of three years, from which they concluded that the characteristics of cloud condensation nuclei were directly determined by the size distribution of aerosol particles and were influenced only to a small extent by their composition.[240] Apparently, it was still not possible to assume that the theoretical principles for the condensation process of cloud droplets to natural aerosols could be taken for granted within the heterogonous community of researchers who addressed questions related to atmospheric research. At least McLaren and Junge considered it necessary to provide an introductory summary on the theoretical basics and, moreover, to add a clear appendix with translations of the variables for their own results. In addition, they included a list of the basic equations for determining the pressure of steam above dissolution droplets along with detailed explanations.[241]

In the early 1970s, in an article for the 1971 MPIC yearbook, Junge defined the "essential tasks of air chemistry research" in terms that could be understood even by the non-specialist public. This field, he wrote, was about the

> "clarification of the cycle [of trace substances, GL] … the type, distribution and yield of sources and sinks associated therewith and the distribution of the components and their average retention time in the atmosphere. These cycles offer a great diversity. The examination of sources and sinks is a very complex interdisciplinary field of work in which collaboration with soil sciences, oceanography and above all biology, especially microbiology, becomes necessary."[242]

Although the continued emphasis of substance cycles already significantly shaped research in the SFB in the 1970s,[243] it took close to a decade and a half for the term biogeochemistry to be accepted as a

240 Junge/McLaren, Relationship.

241 See ibid., 382 and 389.

242 Junge, Stoffkreislauf, 150.

243 Gregor Lax: Interview with Hans-Walter Georgii, April 27, 2015.—Also: Warneck, Geschichte, 3.

label for an interdisciplinary field with explicit emphasis on two Earth spheres. Until the 1980s, journals refused to accept papers submitted as "biogeochemistry research" and in the CPT section of the MPG there was still discussion in 1986 about whether a new department in the MPIC could be named the "Department for Biogeochemistry". In the end, it was permitted.[244]

The second characteristic of the early research in atmospheric chemistry after the start-up of the SFB 73 was the need to promote the further development of methods and instruments in order to even begin to meet the challenges of research on atmospheric trace gases.[245] One clear piece of evidence for this was the budget for appropriate equipment, which amounted to 572,900 DM for the years 1970 – 1972 alone.[246] At the MPIC, the project groups of Seiler and Warneck in particular worked on improving existing methods and instruments as well as new developments. Funding for equipment and components required for prototypes was not always guaranteed by the MPG; DFG funds from the SFB 73 were also used.[247]

Considerable successes soon emerged in the measurement of the expansion and concentration of trace gases in the atmosphere, which resulted in growing recognition in statements by DFG representatives to the SFB 73 as well.[248] One early innovation was a method presented

244 See Andreae, Biogeochemische Forschung, 168 f.

245 Of course, atmospheric chemistry is dependent on a permanent (further) development of its instruments and methods. This in particular holds true for the early years, when the SFB also began its work. Mass spectroscopic and microscopic approaches in particular had to be adapted to the needs of aerosol and trace gas analyses. (Interview Gregor Lax with Hans-Walter Georgii, April 27, 2015).

246 See DFG-Antrag for the years 1970 – 1972, October 10, 1969, Sheet 6, in: BA B227/10803 (Sonderforschungsbereich 73).

247 Examples for this in the early phase of the SFB include a high voltage device model HN 10.000-05 budgeted at 14,319 DM (list of equipment needed in: BA B227/010802, SFB 73) and a light source budgeted at 30,000 DM (October 25, 1972, Georgii to Woll, in: BA B227/011037, SFB 73).

248 See Kienitz to Kirste, June 22, 1972, in: BA B227/0011037 (SFB 73).

by Wolfgang Seiler and meteorologist Ulrich Schmidt in 1970 that allowed a more wide-scale measurement of trace gases. It was based on the reduction of mercury oxide by carbon monoxide (CO) and molecular hydrogen (H_2) and the detection of the resulting mercury vapor by line absorption (resonance layer 2537-A).[249] This method laid the foundation for later work on global distribution and detection of central sinks and sources of CO and H_2. In particular, based on their characteristics as sinks for hydroxyl radicals (OH radicals), these trace gases play an important passive role in the greenhouse effect. In the 1970s it was increasingly becoming accepted that OH radicals are significantly involved in the processes of degradation and production of trace gases in the troposphere and this was an important incentive for including classical chemistry more strongly into the field of activity of atmospheric research initially dominated by meteorologists. In contrast to what was previously assumed in meteorology, only a minimum of the existing OH radicals were destroyed by deposition on aerosols. On the contrary, they formed a central sink for important greenhouse gases, in particular CH_4, but were also a source of formaldehyde (CH_2O), for example.[250]

From the beginning, research at the MPIC and the SFB included the importance of the oceans in the storage as well as the production of aerosols and trace gases. Thus, within the scope of expeditions of the research ship Meteor,[251] measurements of N_2O above and within the North Atlantic Ocean were carried out[252] and in 1972 Junge, Seiler, and others reported on the importance of marine microorganisms

249 Schmidt/Seiler, A new Method.

250 Warneck, Geschichte, 7 ff.

251 This was the second German research ship christened with this name in 1964. The name was to serve as a reminder of the first "Meteor" put into operation in 1925 and entrusted with expeditions until 1938. (See Deutsches Hydrografisches Institut Hamburg, Forschungsschiff Meteor, 17.) It remained in active service until 1985 but was replaced by the third "Meteor" in 1986; this ship is still in active service.

252 Junge/Hahn, N_2O Measurements.

as a source for CO and H_2[253] In the second half of the 1970s, the influence of the land biosphere on the production and absorption of atmospheric CO became another key area of the trace gas research.[254] In 1977, the SFB finally hosted a symposium that comprehensively addressed the role of the biosphere as a source and sink for trace gases. It was named *"The influence of the Biosphäre upon the Atmosphäre [sic!]"*[255] and amongst others Paul Crutzen was one of the speakers. Crutzen later became Junge's successor at the MPIC and in 1995 together with Mario Molina und Frank Sherwood Rowland received the first Nobel Prize for Chemistry that was explicitly bestowed for research on atmospheric chemistry.[256]

From the beginning, a potential anthropogenic component in the perturbation of substance cycles was an aspect of interest in the atmospheric research in West Germany. During the 1970s this interest in anthropogenic issues increased successively. As part of the priority research on trace substances, as early as 1967 a project for the "examination of anthropogenic ozonoids" was envisaged by Hans-Walter Georgii in Frankfurt. It is therefore not surprising that the subsequent SFB 73 expressed interest—that was also explicitly welcomed by the experts—in projects addressing anthropogenic influences that were from the outset reflected in their research objectives.[257] Thus, for example, research was carried out "examining anthropogenic and extraterrestrial aerosols" since 1970.[258] Junge too had already consid-

253 Junge et al., Kohlenmonoxid- und Wasserstoffproduktion.

254 Seiler/Giehl, Influence of Plants; Junge, Stable Isotope Fractionation; Seiler, Influence of the Biosphere.

255 March 9, 1977, Seiler to SFB experts and Wilken for information purposes, in: BA B227/012081 (Sonderforschungsbereich 73).

256 Kant et al., Wissenschaftliche Mitglieder, 317 f.

257 See note de Haar, January 14, 1970, in: BA B227/010802 (Sonderforschungsbereich 73).

258 DFG-Antrag für den SFB 73, submitted by Christian Junge on October 1, 1969 for the years 1970 – 1972, in: BA B227/10803 (Sonderforschungsbereich 73).

ered this topic in the early 1960s[259] and later prepared a series of related publications on substance cycles.[260] Such topics established a direct connection to debates within the international scientific community about the possible significance of human impacts on climate that had gained momentum at that time.[261] In the FRG in the mid-1970s, it was again Junge who summarized the status and the significance of issues surrounding this subject in a comprehensible and high-profile manner. In connection with this issue, on June 18, 1975, at the MPG general meeting, he gave a first, much-noticed talk on *"The Development of the Earth's Atmosphere and Impact of Humans on It"*. The lecture notes were then published in the MPG yearbook. In this talk after a comprehensive explanation of the history of the development of specific trace gas cycles, in particular CO_2, Junge finally stated that

> "there [can hardly be] any doubt that in approximately two generations humans will have a serious problem regarding CO_2. Considering the relatively long period involved—in terms of both the economy as well as the atmosphere-ocean system—serious considerations should commence early enough."[262]

This was formulated 25 years before the Anthropocene concept, according to which mankind is the central cause for climate change, started to become popular.[263]

In the following year, Junge presented another lecture to the German Chemists Society (GDCh) entitled *"The Global Influence on the Composition of the Atmosphere by Mankind"*.[264]

In 1976, *"great importance"* was finally attributed to the question of *"a possible influence on trace gas cycles by humans"* within the SFB 73

259 See Junge, Air Chemistry, 25.

260 Junge, Cycle; Simpson, Man.

261 Singer, Will the World; Rasool/Schneider, Atmospheric Carbon Dioxide; Sawyer, Man-made Carbon Dioxide; Landsberg/Machta, Anthropogenic Pollution.

262 Junge, Entstehung, 45.

263 See Trischler, Anthropocene.

264 See Junge to Vogt, October 14, 1975, in: AMPG, III. Abt., ZA 95, folder 1.

official research program.[265] By then, the largest German project on atmospheric processes could no longer be considered simply basic research in the sense of "pure" science even within the academic world,[266] although admittedly the status of the program as "free research"[267] in the sense of independent choice of research topics and allocation of funds was preserved. At least the federal environmental policy in the FRG was defensive against industrial players who declared environmental protection to be an obstacle to innovation,[268] so the generously financed SFB could continue to focus on the influence of anthropogenic sources without fear of differences with potential third-party sponsors. In the middle of the decade, the objectives and also the rhetoric were adapted in the scope of the SFB 73 research application for the years 1977 – 1979, which stated:

> The aim of air chemistry research today is to explain the existence and the behavior of different components of the atmosphere in a quantitative manner. This also includes answering the question as to how the Earth's atmosphere has developed since its inception. This objective of air chemistry research essentially concerns the pure atmosphere and is thus basic research. However, today it is no longer possible to discount the part originating from anthropogenic air pollution, even in areas with clean air. We consider the development of a description of the behavior of trace gases in the atmosphere that is based on models and from which the global

265 DFG, Jahresbericht 1976, p. 754.

266 By 1945, the term basic research had already been assigned numerous attributes. A point of view that dominated until the end of the 1960s in the science policy discourse among science, economy and politics was decisively shaped by a traditionalist image of science that promoted "pure" independent research—free of political and economic influences. Lax, Lineares Modell, 221 – 264.

267 For the connotations of freedom of research in the FRG emerging in the 1950s and 1960s. See also Lax, Lineares Modell, 221 ff.

268 Müller, Innenwelt, 75 f.

> influence of air pollution can also be calculated to be a long-term objective.[269]

What is interesting here is the clear definition of atmospheric research first as basic research, and finally the limitation of this basic nature with the consideration of human influences. It was with an almost apologetic tone that the inability to continue to discount "anthropogenic air pollution" was mentioned. The characteristic phrasing of a "possible influence" by humans that was still in use at the beginning of the decade had become a fact that could not be argued away.

It is hardly surprising that—albeit with some delay compared to other countries, e.g. the USA—an increasing demand for expertise in the atmospheric sciences arose over the 1970s in the FRG, specifically from political bodies. For example, the Federal Ministry of Food, Agriculture and Forestry asked Junge for an assessment of the potential danger to the atmosphere from nitrogen fertilization.[270] At the international level, EURATOM was one organization that requested an opinion on future research fields.[271] With mutual agreement from both anti-nuclear activists and nuclear power supporters, Georgii served partly as an expert assessing potential environmental impacts from the construction of a nuclear plant and its cooling towers in Wyhl.[272] At the beginning of the 1970s, the anti-nuclear movement in Wyhl and its surroundings was one of the first effective mass protests of the ecological movement in the FRG, even though it was not where the movement directly originated from, as is often misrepresented—there had already been several precursors by that time.[273]

---

269 Fortsetzungsantrag für den SFB 73 for the years 1977 – 1979, in: BA B227/012082.

270 Federal minister of food, agriculture, and forestry to Junge, July 29, 1976, in: AMPG, III. Abt., ZA 95, folder 1.

271 Payrissat to Junge, March 14, 1977, in: AMPG, III. Abt., ZA 95, folder 1.

272 Gregor Lax: Interview with Hans-Walter Georgii, April 27, 2015.

273 See: Engels, Geschichte.

No longer was a cornerstone justifying atmospheric research represented solely by (basic) scientific perspectives; now political and economic viewpoints were important and these would result in a significant boom in atmospheric sciences in the decades that followed. Some of the best known theories that equally considered science, politics, the economy and the public were the nuclear winter (for the development of this computer simulation, Paul Crutzen was elected Scientist of the Year 1984 by the US magazine "Discover"), the role of CFCs in the degradation of atmospheric ozone,[274] the current high-profile concept of the Anthropocene,[275] and the geoengineering approach emphasized in the debate since 2006,[276] to name just a few. The third part of this booklet will address these and other topics in more detail.

Overall, in terms of Junge's department at the MPIC and its driving force in the context of the SFB 73, it can be concluded that the guidelines for the future of atmospheric research were determined for the FRG as a whole. The SFB was the first large-scale project funded by the DFG that addressed the analysis of the chemical composition of the Earth's atmosphere. Interestingly, the SFB program remained predominantly basic-science-oriented and did not directly orient towards pressing environmental policy issues, even though the individual players were certainly involved in the respective topic spectra such as acid rain. Uncharacteristically for a DFG project, the research program was strongly shaped by a Max Planck department from the beginning. The foundation for this was already laid in the mid-1960s with the collaboration between Junge, his former student Georgii and Kurt Bullrich. Under Junge's leadership, atmospheric chemistry as a scientific field was established in name and institutionalized for the first time in the FRG with a relatively well-equipped MPI department.

274 Böschen, Risikogenese; Grundmann, Transnational Environmental Policy; Finally: Brüggemann, Ozonschicht.

275 Crutzen/Stoermer, "Anthropocene"; Crutzen, Geology.

276 Crutzen, Albedo Enhancement.

Junge's successor at the MPI, Paul Crutzen, still considered it necessary to demand greater recognition for atmospheric chemistry as an independent field of atmospheric research in international contexts even at the end of the 1970s,[277] but Junge can certainly be referred to as a pioneer in his field beyond the borders of the FRG.

The research carried out in Junge's department had great influence on the SFB program and made central contributions towards anchoring the key concept of reciprocal processes and influences between atmo-, bio-, and geospheres that remain relevant to this day and which forms a pillar of Earth System research that has become widely established. The topics addressed in the SFB also aroused increasing interest from other institutions even in the 1970s. One example was the MPI for Aeronomy in Lindau (today: MPI for Solar System Research in Göttingen), which, like Peter Warneck's SFB group in the middle of the decade, was measuring the distribution of trace gases in the stratosphere using balloon flights.[278]

The networking among the institutions on both a national and an international level was promoted in the context of the SFB—not least due to pressure based on previous assessments—as was the creation of an environment for junior researchers that had never before existed in the same way. Georgii noted in this context that the funds provided by the SFB, especially in combination with the attractive conditions at the MPIC, made the field more appealing to a growing group of junior researchers.[279] However, this younger generation was certainly no longer primarily focused on weather forecasts, but showed interest in the chemical and physical composition of the atmosphere itself and its

277 Sitzungsprotokoll des Committee on Atmospheric Research Sciences (CAS), February 5, 1979, p. 9, in: AMPG, III. Abt., ZA 125, No. 9.

278 See Max-Planck-Institut für Aeronomie, Jahresbericht, chapter 1.1.

279 Gregor Lax: Interview with Hans-Walter Georgii, April 27, 2015. Georgii describes especially the MPIC as a pilgrimage site for young scientists. Indeed, several young researchers from the Universities of Frankfurt, Mainz and Darmstadt amongst others were employed there in the meantime.

effects on and interactions with other Earth spheres. Furthermore, we have seen that several scientists emerged among the players involved in the SFB who addressed the relevant issues and became important figures in contexts of atmospheric sciences in the FRG in the following decades.

3

ATMOSPHERIC CHEMISTRY AND EARTH SYSTEM RESEARCH UNDER THE LEADERSHIP OF PAUL J. CRUTZEN AND MEINRAT O. ANDREAE, 1980 – 2000

At the beginning of the 1980s, approaches based on Earth Systems began to take on increasing significance among international science communities. The term "Earth System" gained popularity following the initiation of the Global Change Program in 1983 and major international projects associated with it, such as the ICSU[280]-supported World Climate Research Program (WCRP), the "International Geosphere-Biosphere Program" (IGBP) and "Diversitas", a project for biodiversity research. The history of the MPI for Chemistry is closely linked with these developments; this link was, and still is, reflected in the topics addressed and in the immediate institute structures as well as in the institution's involvement in major international projects. The establishment of atmospheric chemistry under the leadership of Christian Junge was the first step in this relationship and was followed

280 International Council of Scientific Unions.

by others. One of the main objectives of the final part of this paper is to explain the development of the MPI into an institute with its main focus on the chemistry of the Earth System. In doing so, the structural changes taking place since the end of the 1970s will be discussed in a first step; following this will be a review of select research fields that made this process, that was accelerated in the 1980s, apparent.

In a separate chapter (3.4), a second development will be examined that was equally characterizing and important for the MPIC. Also in the 1980s, research on the influence of the anthroposphere on global processes between and within the Earth's spheres experienced a boom that ultimately reached its peak with the concept of the Anthropocene proposed by Paul Crutzen and Eugene Stoermer twenty years later. As a guiding principle for this second part, it makes sense to focus particularly on the work of Crutzen and his department since the long-lasting emphasis at the MPIC on anthropogenic influences is closely connected with Crutzen's biography and his most important research interests.

3.1 "Geochemistry in the broadest sense": Restructuring of the MPIC at the end of the 1970s

In 1978, for the first time since the reorganization at the end of the 1960s, extensive changes on both a structural and a personnel level were carried out at the MPIC. In April 1978, Christian Junge had announced to his colleagues his decision to take early retirement that year—"science is now over"[281]—"mainly for private reasons"[282]. After receiving emeritus status, Junge largely retired from the atmospheric sciences although he stayed officially associated with the SFB 73

281 Gregor Lax: Interview with Hans-Walter Georgii, April 27, 2015.

282 Protokoll der Abteilungsbesprechung, April 21, 1978, in: AMPG, III. Abt., ZA 95, folder 8.

until 1985. Heinrich Hintenberger and Hermann Wäffler retired at the same time, so that at management levels, atmospheric chemistry, mass spectroscopy and nuclear physics were equally affected. Only cosmochemistry would remain unchanged under the leadership of Wänke. As we shall see, in contrast to the disastrous developments in the 1960s, the future orientation of the institute was determined early on at this time and the appointments were now much more successful.

In 1976, the CPT section considered that up to that point there had been "two completely independent research institutions" at the MPIC—"namely nuclear physics on the one hand and chemistry of the atmosphere and cosmochemistry on the other hand". Now, the task was "[to] guarantee a uniform objective of the institute".[283] A statement by the MPG senate on November 19 determined that the institute should carry out atmospheric chemistry and "geochemistry in the broadest sense".[284] This finally represented the decision to close the Department for Nuclear Physics rather than continue it following Wäffler's retirement.[285] The statement of the CPT section says something about the perception of the MPIC from the outside: research on atmospheric chemistry and cosmochemistry was seen as a coherent complex, whereas the actual collaboration between Junge's and Wänke's departments had more of a sporadic character. In the sources, there is a growing impression that atmospheric chemistry and cosmochemistry are at least in part not impartially opposite.[286] In addition, contrary to the impression of the CPT section, there were definitely collaborations between the cosmochemistry group and Hermann Wäffler's

283 Sitzungsprotokoll der CPT-Sektion, June 23, 1976, in: AMPG, III. Abt., ZA 95, folder 6.

284 Protokoll der Sitzung des MPG-Senats, November 19, 1976, in: AMPG, II. Abt., Rep. 60, No. 85.SP.

285 An essay by Horst Kant that deals with the split-off of nuclear physics at the MPIC is currently in the planning stages.

286 Carsten Reinhardt/Gregor Lax: Interview with Friedrich Begemann, January 6, 2012.—Gregor Lax: Interview with Heike Tilzer, August 14, 2015.

department, as expressed in clear responses by Hermann Wäffler to questions during the split-off of the Nuclear Physics department. Of the Department for Nuclear Physics, only one working group led by physicist Bernhard Ziegler remained; their complete separation from the MPG met with strong opposition from some scientists at the MPIC. Wänke in particular advocated for not letting Ziegler's group, for which he showed great esteem, out of the MPG's hands and, in case of doubt, to leave it if there was a merger with the Heidelberg MPI.[287] That is exactly what happened and Ziegler remained at the MPIC until 1994.

As of March 1978, Heinrich Hintenberger's Department for Isotope Cosmology was adopted by Wänke's long-standing colleague, Friedrich Begemann. Under the leadership of Begemann, the focus of the department was primarily on mass spectroscopic trace gas analyses and isotope abundance in meteorites. Begemann had completed his studies under Friedrich Houtermans first in Göttingen then in Bern. Subsequently, in 1954, he went to the University of Chicago to work under Willard Libby, who in 1960 was awarded the Nobel Prize for Chemistry for his work on radiocarbon dating.[288] In Chicago, Begemann met the discoverer of deuterium, Harold Urey, and Johannes Geiss who had been discussed as successor to Mattauch at the MPIC at the end of the 1960s (see above). In Chicago, one project Begemann worked on was the measurement of noble gases in meteorites; he was also involved in research on the distribution in the atmosphere and in water cycles of the tritium that had been released in the US hydrogen bomb tests that were initiated in 1952 and had intensified in 1954.[289] During his time in Mainz, he did not continue this work that in principle had already incorporated an Earth Systems approach with

287 See Wänke to Lüst, December 21, 1976, in: AMPG, II. Abt., Rep. 66, No. 854.

288 See in this regard the Nobel Prize speech of Libby: Willard F. Libby: Radiocarbon dating, Nobel Lecture, December 12, 1960, URL: See <https://www.nobelprize.org/uploads/2018/06/libby-lecture.pdf>, status: October 9, 2018.

289 See Begemann/Libby, Continental water balance.

anthropogenic components. As previously mentioned, Begemann had been recruited to the MPIC by Friedrich Paneth and came to the institute in 1957, where he worked closely with Heinrich Wänke. After Paneth's death, both were the driving forces behind the development of cosmochemistry at the MPI after an initial period of considerable difficulties.

In 1979, Albrecht W. Hofmann, a graduate of Brown University in Providence (Rhode Island, USA), was appointed as the head of a newly established Department for Geochemistry. Referring to the intended future focus of the institute on geochemistry, Heinrich Wänke had proposed him to the CPT section in June 1978[290] and he was then recommended to apply in October of the same year.[291] In the meantime, Hofmann had become an assistant at Heidelberg University and had subsequently worked as a postdoc at the Carnegie Institution in Washington, D. C.[292] His main interest was research on the history of the Earth's mantle through analysis of trace substances and isotopes in rocks, especially from the ocean floor. This in turn provided excellent opportunities for connections to the mass spectroscopy work in the departments of Wänke and Begemann.

In 1978, Paul Crutzen was appointed as the successor to Christian Junge as head of the Department for Atmospheric Chemistry[293] and he assumed the position in July 1980.

Crutzen's academic history is rather atypical for a scientific career. According to his own statement, he had wanted to study astronomy after high school and go on to work as a scientist but his grade point

290 Sitzungsprotokoll der CPT-Sektion, June 14, 1978, in: AMPG, III. Abt., ZA 95, folder 6.

291 Sitzungsprotokoll der CPT-Sektion, October 24, 1978, in: AMPG, III. Abt., ZA 95, folder 6.

292 See Kant et al., Wissenschaftliche Mitglieder, 330.

293 Protokoll der 90. Senatssitzung der MPG, June 15, 1978, in: AMPG, II. Abt., Rep. 60, No. 90.SP.

average was not good enough for direct admission to study at Amsterdam University, which meant that he would have had to wait several semesters. Instead, he registered for a civil engineer program at the "Middelbare Technische School" in Amsterdam and, after graduating in 1954, he worked for the Amsterdam Bridge Construction Office for four years—with a temporary interruption for compulsory miliary service. In 1956, he met Finn Terttu Soininen. They married in 1958 and moved to Gaevle in Sweden, where Crutzen took a position as an engineer.[294] However, he was not particularly satisfied with his work and his desire to go into science remained. The opportunity finally arose in 1959, when the Meteorological Institute of Stockholm University (MISU), which was already highly renowned, advertised a position for a programmer, at that time a novel profession the details of which were unknown to Crutzen.[295] Nevertheless, he applied successfully for the position and met Bert Bolin, the head of MISU. Bolin was a pioneer of modern atmospheric research who had called for a more extensive examination of the cycles among the Earth's spheres as early as 1959 (see above). Crutzen started to study meteorology, statistics and mathematics in addition to his employment and had excellent opportunities to work together with atmospheric scientists.[296] As a programmer at MISU, he was involved in the early development of computer-based models especially to determine the ozone distribution in the atmosphere. In the first half of the 1960s, he worked for example with James R. Blankenship,[297] an officer from the U. S. Air Force, who at that time was working on his dissertation at MISU and later on played an important role in the context of the weather satellite-based US Defense Meteorological Satellite Program (DMSP).[298]

294 Carsten Reinhardt/Gregor Lax: Interview with Paul Crutzen, November 17, 2011.
295 Ibid.
296 Ibid.
297 Blankenship/Crutzen, A photochemical model.
298 See Hall, History, 18 f.

After completing his doctoral work in 1968, Crutzen moved to Oxford University on a scholarship until 1971, then returned to MISU and finally moved to NCAR in Boulder to work on the "Upper Atmospheric Program".[299] At this point, he came into closer contact to Christian Junge and his employees in Mainz, since his work on the role of nitrogen oxides in ozone degradation in the atmosphere was directly related to the research interests of the working groups in Mainz, in particular the trace gas research of Wolfgang Seiler and Junge. In 1975, Crutzen wrote an article on the possible effects of carbonyl sulfide (COS) on the atmosphere in interaction with the sulfur aerosol layer discovered by Junge (the Junge layer);[300] he also worked on the importance of laughing gas (nitrous oxide, N_2O) for ozone degradation. In 1975, Crutzen came to Mainz to exchange with the scientists there. Jürgen Hahn and Wolfgang Seiler were two who were particularly interested in N_2O cycles at that time. Hahn had carried out measurements above the North East Atlantic and Seiler worked in the context of the SFB 73 on the cycles of trace gases including carbon monoxide (CO), hydrogen (H_2) and methane (CH_2), as well as N_2O.[301] In December 1975, Crutzen contacted Junge by mail and sent him the draft of his paper on N_2O with a request for a critical opinion.[302] The article was published in "Geophysical Research Letters" the following year.[303] With respect to anthropogenic emitted substances, N_2O possibly counts as the most significant trace gas in terms of the degradation of atmospheric ozone.[304]

299 See Kant et al., Wissenschaftliche Mitglieder, 317.

300 Crutzen, The possible importance of CSO.

301 See Hahn, N_2O Measurements; Seiler, Kreislauf.

302 Crutzen to Junge, December 11, 1975, in: AMPG, III. Abt., ZA 95, folder 1.

303 Crutzen, The possible importance.—The article was published in 1976 but had already been written in 1975. See therefore: Crutzen to Junge, December 11, 1975; Junge to Crutzen October 31, 1975, both in: AMPG, III. Abt., ZA 95, folder 1.

304 See Ravishankara et al., Nitrous Oxide.

Early in the second half of the 1970s, when the need for new appointments became clear, some suggestions for new candidates came directly from the MPIC. Heinrich Wänke, who was at the time acting head of Cosmochemistry, initially advocated his preferred candidate as Junge's successor, oceanographer and Svante Arrhenius's grandson, Gustaf O. Arrhenius (born 1922),[305] who had already before been considered by the CPT section.[306] However, from the beginning, Wänke himself estimated the chances of the appointment as low since Arrhenius had good initial conditions and was deep-seated in San Diego.[307] Wänke then also increasingly advocated Friedrich Begemann as a potential head for the department—as already anticipated with success.

In the name of continuing atmospheric chemistry, Christian Junge proposed Paul Crutzen to the CPT section; according to his estimation *"he is by far the most suitable candidate among the not numerous candidates for this position"*.[308] In 1976, Crutzen had already been included in a shortlist of candidates for a directorship at the MPIC, initially as head of the Department for Geochemistry that was to be established. However, the appointment committee initially withheld this idea from the advisory board of the MPG since one *"board member … is the superior of Crutzen and has not been informed [about the appointment] at all"*.[309] As we have already seen, Albrecht Hofmann assumed the leadership of the Department for Geochemistry. For Atmospheric

305 See Laura Harkewicz: Oral History of Gustaf Olof Svante Arrhenius, April 11, 2006, URL: <http://libraries.ucsd.edu/speccoll/siooralhistories/Arrhenius.pdf>, status May 23, 2018.

306 Sitzungsprotokoll der CPT-Sektion, June 23, 1976, in: AMPG, III. Abt., ZA 95, folder 6.

307 See Sitzungsprotokoll der CPT-Sektion October 25, 1977, in: AMPG, III. Abt., ZA 95, folder 6.

308 Ibid.

309 Note Marsch to the president: Zur ersten Sitzung des Fachbeirats (Bereich Geo- und Kosmochemie, MPIC und MPI für Kernphysik, October 28 and 29, 1976, in: AMPG, II. Abt., Rep. 66, No. 853-2.

Fig. 7: Paul J. Crutzen

Chemistry, the CPT section initially envisaged a double appointment in order to acquire both Paul J. Crutzen as well as Dieter Hans Ehhalt (born 1935) for the MPIC.[310] Crutzen would "be responsible for the theoretical sector and Ehhalt for the experimental sector".[311] Ehhalt had studied mathematics and chemistry at Heidelberg University, graduated in 1963 and moved to NCAR in Boulder in 1964, before he finally returned to Germany to the "Kernforschungsanlage Jülich"

310 See Sitzungsprotokoll der CPT-Sektion, June 14, 1978, in: AMPG, III. Abt., ZA 95, folder 6.—Signed off by the senate: Protokoll der 90. Senatssitzung , June 15, 1978, in: AMPG, II. Abt., Rep. 60, No. 90.SP.

311 Protokoll der Abteilungsbesprechung (Atmosphärenchemie), April 21, 1978, in: AMPG, III. Abt., ZA 95, folder 8.—See also: Protokoll der 89. Senatssitzung der MPG, March 17, 1978, in: AMPG, II. Abt., Rep. 60, No. 89.SP.

(KFA).[312] However, in the end he rejected the call to the MPIC and stayed in Jülich—to the regret of some members of the appointment committee.[313] Thus, Crutzen remained as the only candidate for the first appointment round for the continuation of atmospheric chemistry and was ultimately able to take over the entire division on his own.

The largely positive courses of the appointments to the MPIC since the end of the 1970s clearly suggest that the institute and the MPG as a whole became far more attractive to the international science community than it had been in the late 1960s. Statements in this regard by Crutzen and later also by Andreae confirmed this notion. Crutzen felt that the opportunities that had been offered to him at the MPG in 1978 would never have been available in the US.[314] Andreae, after a tour through the German university landscape in 1984, had initially decided not to return to Germany. There were clearly better opportunities in his field in the US. However, the MPG with the liberties and considerable equipment at the Max Planck Departments was exceptional and Andreae withdrew from a commitment he had already made for a professorship at the University of Washington in Seattle in order to go to Mainz in 1987.[315]

With Paul Crutzen's inauguration in 1980, changes in focus took place in the Department of Atmospheric Chemistry, some significant. Although this did not apply to the entire department under Junge's leadership, Junge himself was primarily an experimental scientist, who preferred to prepare his tax invoices using a slide rule due to his aversion to computers.[316] Indeed, the work in his department as well

312 See Ehhalt's CV in the framework of the online presence of the North Rhine-Westphalian Academy of Sciences and Arts, URL: <http://www.awk.nrw.de/akademie/klassen/naturmedizin/ordentliche-mitglieder/ehhalt-dieter-hans.html>, status May 23, 2018.

313 Gregor Lax: Interview with Hans-Walter Georgii, April 27, 2015.

314 Carsten Reinhardt/Gregor Lax: Interview with Paul Crutzen, November 17, 2011.

315 Gregor Lax: Interview with Meinrat O. Andreae, December 2, 2015 in Mainz.

316 Gregor Lax: Interview with Heike Tilzer, August 14, 2015, in Konstanz.

as in the SFB 73 was primarily shaped by empirical working methods—measurements, sample taking and analyses.[317] With Paul Crutzen, who had entered atmospheric research as a programmer under Bert Bolin in Stockholm, computer modeling and simulation was also intensified at the MPIC. Furthermore, research on the anthropogenic impacts on global climate were carried out and anchored at the institute far more strongly than had been the case under the leadership of Christian Junge. In addition, in the 1980s, the institution underwent fundamental developments towards its current focus as an institute for Earth System chemistry, both in terms of the topics addressed as well as in terms of the institution itself. The paragraphs below will first review in detail the progressive orientation towards Earth System Sciences with special attention to the Department for Biogeochemistry (established at the MPIC in 1987) and the topics of biomass combustion and the CLAW hypothesis. Subsequently, Crutzen's research on anthropogenic influences will be discussed; this work is a recurrent theme throughout his scientific biography starting in the early 1970s prior to his time at the MPI to his more recent contributions, specifically the Anthropocene and the geoengineering debate that emerged in the mid-2000s.

3.2 From atmospheric chemistry to Earth System chemistry

It will become apparent in the paragraphs below that Crutzen was retrospectively one of the early atmospheric scientists who can be referred to as an "Earth System"-scientist. Moreover, his initiative significantly contributed to winning over Meinrat O. Andreae for the Mainz MPI, under whose leadership the Department for Biogeochemistry was estab-

317 Georgii confirmed this in retrospect for the approaches in the SFB as a whole (see Gregor Lax: Interview with Hans-Walter Georgii, April 27, 2015).

lished in 1987 and Earth System research was finally anchored at the institute.

Andreae had studied mineralogy with a geochemical focus initially in Karlsruhe and then in Göttingen, where he completed his degree with geochemist Karl Hans Wedepohl (1925 – 2016). In 1974, Wedepohl introduced him to Edward D. Goldberg (1921 – 2008) at the Scripps Institution of Oceanography at the University of California in La Jolla near San Diego, with whom he then undertook doctoral work. Until the completion of his dissertation in 1978, he studied the distribution of arsenic in the natural environment, starting with the methylation of arsenic in plankton from where it passes through the entire marine food chain.[318] During this research, he also discovered dimethyl sulfide (DMS) production by phytoplankton, which became and remains greatly important for biogeochemical and atmospheric scientific research.[319] Both in terms of his research as well as for science policy, Andreae actively advocated for Earth System approaches and for the establishment of structures that would provide an environment for appropriate research up until this day. He and Paul Crutzen were significantly involved in the foundation of the MPI for Biogeochemistry in Jena (MPIBGC) in 1997, a step that made the field that Andreae had for the first time established on an institutional level within the FRG in 1987 into a fixed component of the German research landscape.[320] The circumstances of the foundation of the MPI-GBC will be examined in the further studies of the GMPG project on the history of Earth System Sciences in the MPG.

318 See Andreae, Biogeochemische Forschung, 158 f.

319 See ibid., 160 f.

320 Gregor Lax: Interview with Meinrat O. Andreae, December 2, 2015, in Mainz.— A concept for a Max Planck Institute for Biogeochemical cycles was proposed by Paul Crutzen in 1994. See Crutzen to the members of the Commission "Neuvorhaben: Atmosphärische Kreisläufe", from June 6, 1994, in: AMPG, II. Abt., Rep. 62, No. 498.

Fig. 8: Meinrat O. Andreae

The term biogeochemistry itself gained in popularity as early as the first half of the 1980s, fanned by increasing numbers of studies on material cycles between the bio-, geo- and atmosphere that had started to rise in the early 1970s. It was natural that he emerged as a scientist linked in particular to studies on substance cycles between the Earth's spheres, specifically the bio-, geo- and atmosphere above all, and was quoted early by renowned scientists. This included Crutzen, for example in an article jointly written with Bert Bolin and Edward Goldberg among others in 1983[321] and in an article in 1985 in an anthology titled "The Biogeochemical Cycling of Sulfur and Nitrogen in the Remote Atmosphere" that was edited by Meinrat Andreae, Henning Rodhe (MISU in Stockholm), Robert. J. Charlson (University of Seattle) and James N. Galloway (University of Virginia).[322] In the FRG, however, the field was first manifested as an institution with the establishment of Andreae's department. Also, Andreae was deeply involved in the foundation of the "Partnership for Earth System Sciences" in 2006, together with Johannes Lelieveld (a student and successor of Crutzen in the Atmospheric Chemistry Department), Martin Claußen (born 1955), Jochem Marotzke (born 1959) from the MPI for Meteorology in Hamburg, and Martin Heimann (born 1949,

321 Bolin et al., Interactions.
322 Crutzen, The cycling.

MPI for Biogeochemistry in Jena).[323] The members of the partnership participate in major international programs, including for example the IGBP and the "Earth System Science Partnership" (ESSP), and are also involved in the work of the IPCC.

At the MPIC, under the leadership of Christian Junge, studies had been carried out that not only addressed phenomena of the atmosphere itself but also focused on the reciprocal interactions and cycles of materials between the atmo-, geo-, bio-, and also anthroposphere (see above). Through the 1980s, priorities arose that made huge strides in bringing the institute nearer to its current profile for Earth System research. Although a detailed presentation of the biogeochemical research fields at the MPIC has already been carried out elsewhere,[324] in the paragraphs below a few specific areas will be highlighted that particularly illustrate the increasing understanding of the Earth as a coherent system.

From the beginning of his scientific career, the exchange between the bio- and atmosphere, as well as the importance of the anthroposphere that is so widely debated today, were paramount for Crutzen. Examples of the areas he worked on include the anthropogenic emission of NO_x and artificially produced CFC that finally became the main factor explaining the ozone hole that had been proposed by Crutzen in 1985 and remains valid to this day. These topics will be addressed separately in the section on anthropogenic influences. In addition, the emission of natural trace gases, including for example CO and H_2 as a result of the oxidation of hydrocarbons from plants, was one of Crutzen's research fields.[325]

The most far-reaching fields arising at the end of the 1970s were undoubtedly those that examined the influence of biomass combustion on the atmosphere; this was a subject that later on played an impor-

323 Andreae/Marotzke/Heimann, Partnerschaft Erdsystemforschung.

324 See Andreae, Biogeochemische Forschung.

325 See Zimmermann et al., Estimates.

tant role at the MPIC in Crutzen's as well as in Meinrat O. Andreae's department. The history of biomass combustion equally combines anthropogenic activities (for close to 40,000 years), natural phenomena, changes and adjustment processes in the biosphere, biodiversity and influences on regional and global climate.[326]

As early as 1979 and 1980, working with Junge's former employee Wolfgang Seiler and others, Crutzen studied the combustion of biomass as a source of a number of carbon and nitrogen compounds and for the first time presented estimates of the corresponding emissions[327] that later on were further refined by Crutzen and Andreae amongst others.[328]

Also, around 1980, when he was still an assistant professor at Florida State University in Tallahassee, in the framework of an expedition with the German research ship "Meteor", Andreae found that emissions from biomass burning on the African and Latin American continent drift far out into the equatorial Atlantic. Stimulated by discussions with Paul Crutzen at NCAR in Boulder, these observations increased Andreae's interest in this relatively novel subject area[329] and also paved the way for the subsequent collaboration between what was later the Department for Biogeochemistry and Crutzen's Atmospheric Chemistry Department at the MPIC.

As early as 1984, Crutzen and Andreae had jointly worked on a paper on the role of atmospheric chemistry in which ideas for future research were summarized, some of which became a part of the "International Geosphere-Biosphere Program" (IGBP) starting in 1987. The text appeared in the "Global Change" anthology supported by the ICSU in 1985, which also included the discovery of the "ozone

326 See the article in the overview—and memoranda—like anthology of Levine: Andreae, Biomass burning.

327 Crutzen/Meinrat, Biomass Burning; Seiler/Crutzen, Estimates.

328 See Crutzen/Andreae, Biomass Burning.

329 See Andreae, Biogeochemische Forschung, 164.

Fig. 9: Research Vessel "Meteor" from 1964

hole".[330] With their work on biomass combustion, Andreae and Crutzen were heavily involved in major projects under the umbrella of the "International Global Atmospheric Chemistry Programme" (IGAC) that was affiliated with the IGBP. The five key objectives of IGAC included research on global distributions and trends of relevant substances in the atmosphere, exchange processes above the biosphere, transformation processes in the gas phase, the creation of a stronger theoretical basis for modeling and prediction instruments and the examination of multiphase processes.[331] Much later, in 2012, this last area was firmly anchored at the MPIC in the form of the Department for Multiphase Chemistry under the Leadership of Ulrich Pöschl. The front cover of the first publicly accessible status report of the IGAC of 1989 already spoke volumes regarding the importance of biomass

330 Crutzen/Andreae, Atmospheric Chemistry.

331 See Galbally: International Global Atmospheric Chemistry, 45.

combustion as the object of research at the time. It showed an image of emission measurements during a forest fire in Brazil in 1979 that had been made available by Crutzen and P. Zimmermann.

After taking this photo, it would take almost another ten years before projects were targeted on a large scale to examine the importance of biomass combustion for global atmospheric processes. For a long time, the scientific community refused to believe the actual extent of the impacts of certain phenomena over long distances. In this context, Andreae describes in detail how the effects of biomass combustion had gradually emerged more and more obviously through the 1980s, mainly as a side effort in projects with other focuses. This included the discovery by Andreae in the equatorial Atlantic on the

Fig. 10: Emission measurement during a fire in the rainforest of Brazil, 1979

"Meteor" trip in 1980 that was mentioned earlier and other insights from the mid-1980s. Examples include the "Amazon Boundary Layer Experiments" (ABLE-2A) carried out in 1985, the measurement campaigns in the rainforest of the Congo in Central Africa in 1988, and the uncovering of further evidence of the presence of masses of fumes above the Atlantic resulting from the "Chemical Instrumentation Test and Evaluation Expedition (CITE-3)" carried out by NASA in 1989.[332] It appears that a consolidation of all the evidence collected almost by chance in the framework of these efforts finally legitimized initiatives that were explicitly focused on examining the effects of biomass combustion. In the scope of the first two of these projects, two different savanna types were studied on the African continent. In the "Fire of Savanna/Dynamique et Chimie Atmosphèrique en Forêt Equatoriale" (FOS/DECAFE-91) project that was set up on a small scale in 1991, emissions of CO, CO_2 and NO_x were measured in the "moist" savanna where there is a relatively high proportion of biomass per hectare and approx. 80 % moisture content in Lamto (Guinea) in Ivory Coast. In the following year, corresponding measurements were then carried out in the scope of the "Southern African Fire-Atmosphere Research Initiative" (SAFARI-92) in whose context the "dry" savanna was investigated in Kruger National Park in South Africa, which is characterized by a relatively low proportion of biomass per hectare and a moisture content of approx. 10 – 20 %.[333] Emissions of intentionally lit fires at these locations were measured directly at ground level as well as over long air routes over the African continent.[334] In the mid-1990s, in the course of other undertakings, additional measurements were carried out on the "breathing" as well as the of burning biomass above the Latin American continent. This included in particular the

332 Andreae, Biogeochemische Forschung, 175 ff.

333 See: Lacaux et al., NO_x emissions.—Authors of this article include Thomas Kuhlbusch among others who were in Crutzen's department at that time.

334 See Andreae: Biogeochemische Forschung, 176.

"Cooperative LBA Airborne Regional Experiment" (CLAIRE-98, continued with CLAIRE-2001) that was initiated in 1998 and was named by Andreae (in honor of his daughter Claire M. Andreae); it was this effort in particular that provided the first proof that emissions from vegetation fires have the potential to rise very far into the tropopause.[335]

The examination of biomass combustion equally links different sections of the Earth System and the anthropogenic sphere in a way unlike any other field. In particular, the historicity of the encounter of humans with fire as a natural threat on the one hand and as a purposefully deployed "tool" on the other hand has existed for several millennia and has still not lost its actuality today. The massive fires in California in recent times[336] and the enormous extent of slash-and-burn practice in Indonesia are just two among several possible examples. Appropriate reactions in politics and science became increasingly apparent at both national and international level starting in the mid-1990s. For Germany, the Global Fire Monitoring Center (GFMC) is a good example; it was initiated in 1998 as a branch of Andreae's Department of Biogeochemistry at the MPIC under the leadership of Johannes Goldammer (born 1949) and was located at the University of Freiburg. The GFMC is closely linked to the "Wildland Fire Advisory Group", which originated in the framework of the UN International Strategy for Disaster Reduction (UNISDR), and the "Global Wildland Fire Network (GWFN)".[337]

335 See Andreae et al., Transport.

336 See e. g. Zeit Online: Die Feuer verändern sich im Minutentakt, October 23, 2017, <https://www.zeit.de/gesellschaft/zeitgeschehen/2017-10/kalifornien-waldbraende-san-francisco-tote-suche-vermisste>, status: October 10, 2018.—Christiane Heil: Mutter Natur entscheidet, wann wir löschen können, FAZ Online, December 6, 2017, <http://www.faz.net/aktuell/gesellschaft/ungluecke/d-15327788.html>, status: October 10, 2018.

337 See GFMC website: URL: <http://gfmc.online/intro/About1.html>, status: October 10, 2018.

3.3 The CLAW hypothesis: Research on the basis of an "Earth System theory"

In 1987, the year of his inauguration as director of the Department of Biogeochemistry at the MPIC, and together with Robert Charlson (born 1936), James Lovelock (born 1919) and Steve Warren (born 1945), Meinrat O. Andreae published the essay "Oceanic phytoplankton, atmospheric sulfur, cloud albedo and climate" in the journal "Nature". With more than 2400 (recorded) citations already to date,[338] the article has received huge attention in research and the key statements in it have since become known as the "CLAW hypothesis". The name CLAW derives from the last names of the four authors and the approach can be understood only against the background of the Gaia hypothesis. This hypothesis was first proposed by atmospheric scientist James Lovelock at the beginning of the 1970s[339] and has been further developed, in particular, in collaboration with biologist Lynn Margulis (1938 – 2011) in the following years.[340] The Gaia hypothesis is the first approach discussed in detail in the international research that was clearly based on an understanding of chemical-physical processes in the Earth System and was to serve as a heuristic thought structure for a broad range of research in the natural sciences. Inspired by a cybernetic system understanding, the hypothesis, named after the Greek Earth goddess, is based on the fundamental assumption that the biosphere regulates itself as a kind of superorganism and continuously keeps the required climatic conditions for earthly life in balance.[341]

Using the computer simulation "Daisyworld" and together with Andrew J. Watson (born 1952) from the "Marine Biological Associa-

338 The Web of Science Index showed 429 citations on November 26, 2015.

339 See: Lovelock, Gaia (1972).

340 See preface in: Lovelock, Gaia (1979), 2 f.

341 For the borrowing of the system term from cybernetics, see: Lovelock, Gaia (1979), Chap. 4.

Fig. 11: Authors of the CLAW hypothesis: Robert Charlson, James Lovelock, Meinrat O. Andreae, Steve Warren (from left)

tion" in Plymouth, England, Lovelock suggested that a fictitious planet circling around a star similar to our sun with a linearly increasing heat curve could hold its climate relatively constant over a longer period only if there are two forms of life on it that each respond differently to sunlight.[342] Lovelock and Watson used two different species of daisies for their experiment. The black one absorbed light and warmed up the planet while the white one, like the Earth's albedo, reflected light and thus contributed to cooling. According to the simulation, the more solar radiation warmed up the planet, the more white and fewer black daisies grew on the surface and held the temperature constant until the entire planet was populated by white daisies and finally there was nothing to oppose the continually increasing solar heat curve.[343]

342 See Watson/Lovelock, Biological homeostasis.

343 See ibid., 286 f.

"Gaia" was discussed in detail in the 1980s and continues to this day to be pursued at times, partly criticized and yet also mystified and personified outside of science. The Gaia picture series by English artist Josephine Wall can be used as a current example for the latter in which the Earth goddess Gaia creates Earth and oceans.[344] The Gaia hypothesis and the Daisyworld thought experiment raised the question for research about which mechanisms would be relevant for self-regulation of the Earth's biosphere and it was precisely here that the CLAW hypothesis started with an initial serious proposal. This consisted of a feedback effect in which the production of dimethyl sulfide (DMS) by specific phytoplankton species has a key role: the greater the solar radiation above the Earth's oceans, the more phytoplankton there is that emits DMS and hence there is more DMS produced and emitted into the marine troposphere. The DMS in turn has a major role in the formation of cloud condensation nuclei and thus the formation of clouds above the oceans. The Earth's albedo increases, solar radiation decreases limiting algae production, and as a final consequence the production and release of DMS into the troposphere is also expected to be limited. However, the attempt to measure the concentration of DMS using satellites that detect phytoplankton turned out to be highly complex due to the wide range of starting conditions for the production of DMS. To this day, no process-based model has been developed that can provide an accurate description of the release of oceanic DMS into the atmosphere.[345] After some time, Andreae distanced himself from the CLAW hypothesis that was marked by chemical and mechanical approaches. The large number of very divergent plankton species with different characteristics, specializations, growth rates etc. has created a highly complex picture the biological dimensions

344 See the presentations of Josephine Wall on the revival of the Earth by Gaia for example. URL: <http://josephinewall.co.uk/art-gallery/goddesses/>, Status: September 19, 2018.

345 See Andreae, Biogeochemische Forschung, 171.

and relationships of which remain to be fully understood.[346] *"What remains [from the CLAW hypothesis]"*, Andreae states in retrospect

> "is ultimately that biology produces volatile sulfur compounds in the ocean that merge into the atmosphere via cross connections and there have an effect on physical processes in the clouds and influence the climate. How the climate in turn influences the marine biology and in which direction the DMS production is driven—this is where I see a lot of questions."[347]

Research on the DMS cycle remains a current research field that today is increasingly also focused on the presence of DMS in the land biosphere.[348]

3.4 From the examination of anthropogenic influences to the "Anthropocene"

"Stop it! We are no longer in the Holocene, we are in the Anthropocene," with these words, Paul Crutzen remembers in retrospect a meeting of the IGBP in Cuernavaca (Mexico) in 2000. He addressed a speaker there with these words in whose talk the term "Holocene", at that time common for the current climatic era, was used repeatedly.[349] Crutzen could not have known then that with this interjection he would give a name to a potentially new geological era in which humans will be the decisive climatic factor: the Anthropocene. In 2000, together with Eugene Stoermer, Crutzen published an article barely two pages in

346 Gregor Lax: Interview with Meinrat Andreae, December 2, 2015.

347 Ibid.

348 Recently: Jardine et al., Dimethyl sulfide.—Six of the authors belong to the Department for Biogeochemistry (Andreae, Kesselmeier, Williams, Behrendt, Veres und Derstroff).

349 See Carsten Reinhardt/Gregor Lax: Interview with Paul Crutzen, November 17, 2011.

length that outlined in greater detail the "Anthropocene"—still presented in quotes at the time. As indicators of this era, the authors pointed to the global accumulation of human population and urbanization, the consumption of fossil fuels and water, the increase of synthetic chemicals, overfishing, and the loss of biodiversity most notably in tropical regions, the massive increase of greenhouse gases and other harmful gaseous substances (esp. CO_2, CH_4, NO and SO_2), the long-term evidence of anthropogenic influences in lakes and the increase of natural disasters as well as the risk of man-made disasters, in particular the "Nuclear Winter",[350] which will be discussed below.

In 2009, the International Union of Geological Sciences (IUGS) reacted to the Anthropocene proposal and deployed the "Anthropocene" Working Group. This group was established within the Subcommission for Quaternary Stratigraphy, which itself was a part of the "International Commission on Stratigraphy" (ICS) at the IUGS. The objectives formulated by the working group were *"to examine the status, hierarchical level and definition of the Anthropocene as a potential new formal division of the Geological Time Scale"*.[351] The commission, however, includes solely proponents of an Anthropocene—including the chairman Jan Zalasiewicz. The scientific spectrum of the members is concentrated in climate and Earth System sciences, but occasionally also with other research areas, e.g. the history of sciences, which are represented by Naomi Oreskes (born 1958).[352] Against this background it is unsurprising that the working group came to the conclusion that the Anthropocene has been reached. But it has not

350 Crutzen/Stoermer, "Anthropocene".—The term became more popular following the often-quoted essay: Crutzen, Geology.

351 Anthropocene Working Group of the Subcommission on Qarternary Stratigraphy (International Commission on Stratigraphy), Newsletter No. 1 (2009), p. 1, URL: <http://quaternary.stratigraphy.org/workinggroups/anthropo/Anthropnewsl1.doc>, status: May 23, 2018.

352 See the list of the Commission members: URL: <http://quaternary.stratigraphy.org/ working-groups/anthropocene/>. Status: October 10, 2018.

been finally clarified whether it started after the Holocene or as part of the Holocene.[353]

The idea of a new Earth era widely triggered debates not only in geo- and atmospheric sciences but also in social sciences and humanities, which have increasingly participated in the discussion since the 2010s. The latter not only made a valid joint claim as to the definition and conceptualization of the new era, but also asked about the basic consequences of the Anthropocene. For example, to what extent does the concept of the Anthropos need to be revised,[354] whether the relationship between humans and nature ought to or could be rethought[355] and whether the term Anthropocene is perhaps misleading because it could suggest that humans, as the main factor influencing the climate overall, do control, could control, or should control the Earth System.[356]

By now, the Anthropocene was anticipated in the media as well. This included public mass media that had developed a growing interest over the past years and in which some science journalists had already shown an interest in the Anthropocene.[357] In addition, the Anthropocene had become a fixed component of exhibition and educational institutions, some highly prominent, for example in the context of the "Haus der

353 See online article of the committee member Ellis: Erle Ellis (2013): Anthropocene, <https://editors.eol.org/eoearth/wiki/Anthropocene>, status: October 10, 2018.

354 See Palsson et al., Reconceptualizing.

355 See Dalby et al., After the Anthropocene.

356 See Malm/Hornburg, geology.

357 For an example, see the work of Bojanowski and Schwägerl. To name only some readily accessible examples: Axel Bojanowski/Christian Schwägerl: Debatte um neues Erdzeitalter: was vom Menschen übrig bleibt, in: Spiegel Online, July 4, 2011, URL: <http://www.spiegel.de/wissenschaft/natur/a-769581.html>, status May 23, 2018.—Christian Schwägerl: Planet der Menschen, in: Zeit Online, February 18, 2014, URL: <http://www.zeit.de/zeit-wissen/2014/02/anthropozaen-planet-der-menschen>, status: May 23, 2018.—Axel Bojanowski: Debatte über Anthropozän: Forscher präsentieren Beweise für neues Menschenzeitalter, in: Spiegel online, August 25, 2014, URL: <http://www.spiegel.de/wissenschaft/natur/a-987349.html>, status: May 23, 2018.

Kulturen der Welt" ("House of the World's Cultures") exhibition in Berlin, or in the Deutsches Museum in Munich. The former initiated the "Anthropocene project" in 2013/14 with the participation of the Berlin Max Planck Institute for the History of Science among others.[358] In 2016, the Deutsches Museum launched the exhibition "Welcome to the Anthropocene. Our Responsibility for the Future of the Earth".[359]

The concept of the "Anthropocene" marks the culmination of an occupation with the influence of humans on climate and environment that had extended throughout Crutzen's scientific career. This was reflected in the establishment of several research focuses in the Department for Atmospheric Chemistry at the MPIC. Under Crutzen's leadership, anthropogenic influences had played a greater role than before. Despite the fact that he dealt with politically charged topics such as "acid rain",[360] Christian Junge had considered himself more of a scientist conducting basic research and, together with Georgii and Bullrich, in the mid-1970s he still defined the tasks of the SFB 73 as primarily "pure research", although those involved were aware of the considerable political potential of studying anthropogenic influences (see above). Crutzen, in contrast, at an early stage saw it as his task to influence political decision-making processes as an active voice speaking on the basis of scientific knowledge. In this regard, he differs considerably from Junge, who appeared far less offensive and withdrew entirely from the field of atmospheric research after his retirement. In 2011, Crutzen stated in retrospect:

> "I don't carry out 'pure science', although originally that was my goal. Until I discovered … this is more than science, since humans

358 See homepage of the Haus der Kulturen der Welt. URL: <http://www.hkw.de/de/programm/projekte/2014/anthropozaen/anthropozaen_2013_2014.php>, status: May 23, 2018.

359 See homepage of the Deutsches Museum München: <http://www.deutsches-museum.de/ausstellungen/sonderausstellungen/rueckblick/2015/anthropozaen/>, status May 23, 2018.

360 See e. g. Junge/Werby, Concentration; Junge, Chemical Composition.

are involved. And this was an important part of my research … I discovered then that nitrogen oxides influence the ozone and even the climate."[361]

In fact, an interesting pattern can be seen in Crutzen's work over long stretches from the beginning of his scientific career: Crutzen's basic research was directly carried over to specific sociopolitical or environmental issues—frequently by himself. These issues in turn were often actively introduced by him in the sense of a compass in the contexts of political decision making. Several of Crutzen's key research areas during and before his time at the MPIC dealt in different ways with the anthropogenic role in climate and the Earth System. In 1968, he had shown in his dissertation that the status of the theory at that time could not adequately explain the ozone distribution in the stratosphere; this applied in particular at heights of 30 – 35 km.[362] He worked to determine numeric parameters that would allow more precise statements regarding O_3 distribution and soon after discovered the eminent importance of nitrogen oxides (NO_X) in the degradation of stratospheric ozone by catalysis.[363] Early on, this work directed his interest towards the global effects of anthropogenic influences which at the time were receiving little attention. In the dedication to his wife in one of his first papers on NO_X, he had anticipatorily written *"I hope this will not disturb our lives too much"*.[364]

In the following chapters, a few themes relating to anthropogenic influences will be explained in more detail. We start approximately one decade prior to Crutzen's accession to office, indeed at the beginning of his scientific carreer, when he began to address the effect of NO_X in the atmosphere and the influence of emissions from airplanes.

361 Carsten Reinhardt/Gregor Lax: Interview with Paul Crutzen, November 17, 2011.

362 See Crutzen, Determination.—Both the approach of Chapman, already deemed "classical theory", as well as Hampson's work from 1965/66 were the focus (see: Chapman, theory and Hampson, Chemiluminescent emissions).

363 See Crutzen, The influence of nitrogen oxides.

364 Carsten Reinhardt/Gregor Lax: Interview with Paul Crutzen, November 17, 2011.

3.4.1 Influences of air traffic on the atmosphere

Among Crutzen's earlier areas of research was an interest in the possible effects of air traffic on the atmosphere, something he had studied since the early 1970s. This topic was significant in particular in the context of supersonic flights that had attracted the public's attention since the end of the 1960s; the medium-term goal was to introduce such flights in civil aviation as well. The first successful test flight of a supersonic passenger plane involved a Soviet Tupolev TU-144 at the end of 1968; this plane was sometimes referred to as the "Concordski" by Western media due to its striking similarity to the French Concorde. It took off before the Concorde test flight that was carried out a couple of weeks later.[365] The "Super Sonic Transports" (SSTs) were associated with a number of different challenges that gave rise to repeated criticism in the following decades: they were very expensive, very loud and there was reason to believe that the emission of NO_x into the stratosphere could significantly affect the ozone layer. Several organizations against SSTs were founded, including the "Anti-Concorde Project" in England, the "Citizen's League against the Sonic boom", the "Coalition against the SST" in the US and the "Europäische Vereinigung gegen die schädlichen Auswirkungen des Luftverkehrs" ("European Association Against the Harmful Effects of Air Traffic") located in Frankfurt am Main and which was a little less specific to a single country and more general in focus. Several countries had already announced that they would not open their airspace to supersonic flights, including Canada, large parts of Scandinavia, the Netherlands and Switzerland.[366] In 1971, Crutzen was involved in the first conference of the "Climate Impact Assessment Program" (CIAP) Bureau of the U.S. Department

365 See the description of the exhibited original copy of the TU-144 as part of the website of the technical museums Sinsheim and Speyer, URL: <http://www.bredow-web.de/Sinsheim-Speyer/Tupolev_TU_144/tupolev_tu_144.html>, status: May 23, 2018.

366 See Crutzen, SST's, 43.

of Transportation that had been established in connection with the SST issue. His contribution addressed the possible effects of NO_x emitted by SSTs on the stratosphere.[367] In his opinion it was clear that insufficient attention was being given to environmental problems in general. In 1972, he published an article in which he stated that the criticism of SSTs until then had been dominated primarily by economic and political objections that have already been briefly outlined, while a potentially alarming impact on the ozone layer had only been discussed on the side.[368] The article explained in detail that the SSTs planned by Great Britain, the US and the USSR in the middle tropopause could cause a substantial increase of NO_x in the stratosphere. Below 40 km, where the ozone layer is densest, these flights would result in massive destruction of O_3 molecules by catalytic reactions.[369] It was feared that a fleet of 500 regularly scheduled supersonic planes could reduce the ozone layer by half, or perhaps even completely.[370] However, precise estimations were actually not possible at the time, since research on ozone degradation processes in the atmosphere was still in its infancy.[371] In light of technical developments in aircraft construction and other innovations, more recent estimates show approximately 1 % damage.[372]

Ultimately, the criticism at the time did not prevent the use of SSTs in civil aviation. In 1976, the Concorde was finally placed into service as the first ultrasonic passenger plane and remained in service until 2003. For economic reasons, however, the number of jets remained very limited and in 2003, for safety reasons, the Concorde suffered

367 See Crutzen, Photochemistry.

368 See Crutzen, SST's, 41.

369 See ibid., esp. p. 42 and 46.

370 See ibid., 41 f.

371 A summary of the research that had intensified until the 1990s can be found in Stolarski et al.: 1995 Scientific Assessment.

372 See Houghton et al., Climate Change 1995, p. 96.

the same fate as the Tupolev in 1978: the crash of Paris in July 2000, which was covered in detail by the media, resulted in the termination of scheduled Concorde flights.[373] Nevertheless, investigations of the effects of aviation remained a topic in atmospheric research as a whole. In the 1980s, as director at the MPIC, Crutzen initially focused on other fields and it was not until the early 1990s that he once again addressed questions about the impacts of aviation on the atmosphere.[374]

3.4.2 NO_x, CFCs and the discovery of the ozone hole

Another major field of research in the 1970s was the effects of natural and artificial propellants on the ozone layer. In 1974, Mario Molina (born 1943) and Frank Sherwood Rowland (1927 – 2012) suggested that industrially produced chlorofluorocarbons (CFCs), which do not exist naturally, could play a significant role in ozone degradation. In their well-known article in the journal "Nature", they pointed out the dangers of chlorofluoromethane (HCFC-31), which first, was being used to an ever greater extent and thus would reach the atmosphere and second, had an expected residence time in the atmosphere of between 40 and 150 years.[375] Molina and Rowland went on to note that as the proportion of HCFC-31 grew, the proportion of chlorine atoms in the stratosphere would also increase sharply. The catalytic reaction between chlorine and O_3 molecules would result in rapid degradation of the ozone layer. Molina and Rowland listed NO_x as a comparison

373 See Spiegel cover story: Pott et al., Richtung Zukunft.—See also: Deckstein, Think small.—Michael Klaesgen: Wilde Jagd am Himmel. In Paris zerschellte der Mythos der Sicherheit, in: Zeit No. 31/2000, here from: Zeit online archive, URL: <https://www.zeit.de/2000/31/Wilde_Jagd_am_Himmel>, status: May 23, 2018.

374 Amongst others: Brühl et al., Increase; Fischer et al., Observations.—In the context of international events for example: Brühl/Crutzen, The atmospheric chemical effects; Grooß et al., Influence.

375 See Frank Rowland/Molina, Stratospheric sink, 810.

for this catalytic reaction chain and cited amongst others Crutzen's work from 1971.[376] Also in 1974, Crutzen himself pointed out a possible ozone reduction by CFCs, in particular by CF_2Cl_2 and $CFCl_3$,[377] and wrote an article in the same year that summarized the potential anthropogenic influences on the degradation of atmospheric ozone and estimated an overall rate that until then had been debated. CFCs, SSTs and the global production of NO_x were all included.[378] NO_x emissions in the event of a possible nuclear war were also alluded to.[379] In September 1975, Rowland and Crutzen gave talks on the subject of CFCs at the conference of the World Meteorological Organization (WMO).[380]

The political explosiveness of the work by Molina, Rowland and Crutzen, as well as others, was based on the fact that CFCs were almost irreplaceable substances at the time, with industrial applications ranging from coolant production to spray can propellants. Thus, these substances were of central importance to the large-scale chemical industry around the world.[381] Public discussion of the possibility of restricting CFC production was initially more reserved in the FRG than in the United States, for example, where heated debate had started at a relatively early stage. The affected industry in the Federal Republic of Germany initially argued that no solid evidence could be produced to support the hypothesis of ozone degradation by CFCs at

376 See ibid.

377 See Crutzen, Estimates of possible future ozone reductions.

378 See Crutzen, Estimates of Possible Variation, 201.

379 See ibid., here 206 f.—This hypothesis was proposed by John Hampson in the same year. See: Hampson, Photochemical war.

380 See AMPG, III. Abt., ZA 125, No. 9.

381 In scientific research, several studies have been presented on the social debates around CFCs, in particular for the FRG and the US. Mentioned here: Böschen, Risikogenese; Grundmann, Transnational Environmental Policy.

such an alarming magnitude; they also highlighted the elimination of innumerable jobs if production were to stop.[382]

Even so, although somewhat later than in the US, a more profound change in policy occurred with the discovery of the Antarctic ozone hole in 1985, which became famous by the publication "Large Losses of Total Ozone in Antarctica Reveal Seasonal ClO_x/NO_x Interaction" by Joe Farman, Brian Gardiner and Jonathan Shanklin.[383] But it was Paul Crutzen and Frank Arnold from the MPI for Nuclear Physics in Heidelberg who provided the underlying explanation shortly after this discovery:[384] in the darkness of the polar winter, a cold air vortex is formed that supports the formation of polar stratospheric clouds (PSCs). These clouds consist to a large extent of acid molecules (primarily nitric acid) and are formed in the aerosol veil that had been discovered by Christian Junge at the beginning of the 1960s (Junge layer, see earlier). Chlorine and bromine molecules are deposited on the surface of the particles in the PSCs.[385] At the end of the polar winter, increasing solar radiation means that the deposited molecules are activated in photochemical reactions. Then a catalytic reaction occurs, in which O_3 molecules are degraded so rapidly that a veritable hole is created in the ozone layer.

In the Department for Atmospheric Chemistry, anthropogenic influences on the ozone layer remained a fixed component of research even in the years that followed.[386] Crutzen's employee Christoph Brühl was particularly involved in the calculations of the ozone hole that

382 See Brüggemann, Ozonschicht, 175.

383 Farman et al., Large Losses.

384 Crutzen/Arnold, Nitric acid.

385 The features of chlorine and bromine that destroy the ozone layer were a focus of Crutzen's department from the beginning. Amongst others: Berg et al., First measurements; Gidel et al., A two-dimensional photochemical model.

386 See: Crutzen/Arnold, Nitric acid; Barrie et al., Ozone destruction; Crutzen et al., Nitric acid haze formation.

Fig. 12: Polar Stratospheric Clouds, NASA 2005

were carried out by the MPIC until well into the 1990s.[387] Brühl had studied meteorology at Mainz University, worked at the MPIC, and graduated in 1987. In the meantime, he worked at NCAR in Boulder and after finishing his doctorate he entered the working group for computer modeling in Crutzen's department.[388]

With the discovery of and explanation for the ozone hole, CFCs entered public awareness as a man-made "ozone killer" and international response on both economic and political levels followed, although at times delayed depending on the region. Whereas CFC

387 For example: Brühl/Crutzen, Scenarios; Crutzen et al., Nitric acid haze formation; Brühl/Crutzen, Ozone; Brühl et al., Increase; Crutzen et al., On the potential importance.

388 E-mail Christoph Brühl to Gregor Lax, November 9, 2015.

producer DuPont in the US initiated a relatively quick response and made considerable R&D efforts towards the development of alternative substances, a reorientation at Hoechst AG and at Kali Chemie AG (which closed in 2011) was not carried out until 1986/87.[389] Nevertheless, the ozone hole had caused a sharp increase in public interest in climate-related topics in general and in the production of CFCs in particular. The position of proponents for regulation of CFC production was also strengthened in the framework of federal policy in Western Germany. Despite the somewhat delayed reaction by the German industry compared to the US, on a political level the FRG in 1987 finally became one of the driving forces in Europe advocating the establishment of the Montreal Protocol that restricts the production of CFCs on a global scale.[390]

For their pioneering work on NO_x and CFCs and later studies on the ozone hole, Crutzen, Molina and Rowland jointly won the first Nobel Prize for Chemistry in 1995 that was decidedly awarded for research on atmospheric chemistry.[391] Jaenicke considers this a final acknowledgment of this department that had been introduced by Christian Junge at the MPIC in 1968.[392]

3.4.3 Studies on the "Nuclear Winter"

The first work on the hypothesis of a Nuclear Winter arose at the beginning of Crutzen's term as director at the MPIC; this topic played a significant role in both scientific and public discourse of the 1980s.[393] The hypothesis predicted a long-term obfuscation of the Earth's atmo-

---

389 See Brüggemann, Ozonschicht, 184ff.

390 See ibid., 181.

391 See Press release of the Nobel Prize Organization, „The Nobel Prize in Chemistry 1995“: <https://www.nobelprize.org/prizes/chemistry/1995/press-release/>, status: October 9, 2018.

392 See Jaenicke, Erfindung, 187.

393 See Badash's work: Badash, Nuclear Winter's Tale.

sphere caused by the extreme dust formation expected in the event of a nuclear war. Darkness, cold and subsequent poor harvests and famine around the globe would result. Thirty years later, Crutzen retrospectively referred to this idea as *"probably ... the most important I ever had"*.[394]

Potential after-effects of the use of nuclear weapons on the Earth's atmosphere had already been discussed in the mid-1970s. Atmospheric researcher John Hampson, a professor at Laval University in Quebec, Canada at the time, had pointed out the danger of massive release of NO_x in the event of nuclear weapons use. At the time, Hampson felt that people were far from understanding the atmosphere, but a nuclear war could result in a massive thinning of the ozone layer through the well-known photochemical reaction of NO_x with O_3,[395] Initially, there was little focus on the question as to whether the use of nuclear weapons could also cause long-term climatic cooling due to the ensuing formation of dust, resulting from bombshells and the following fires. This changed in 1982, when Crutzen and John William Birks published the article "The atmosphere after a nuclear war: Twilight at noon" in *Ambio*.[396] In addition to the expected high rate of NO_2 emission into the atmosphere, the authors emphasized the significance of the formation of smoke from extensive fires expected in cities, forests and oil and gas fields following a nuclear exchange. Thus, the focus of the approach was the consequence of the burning of biomass and material, which Crutzen had already started to work on before his time as MPI director. Like his earlier work on NO_x, these approaches were "applied" to a certain extent in the context of the Nuclear Winter, which, against the background of the Cold War, was

394 Carsten Reinhardt/Gregor Lax: Interview with Paul Crutzen, November 17, 2011. See also the published extract of the interview in: Crutzen/Lax/Reinhardt, Paul Crutzen, 49.

395 See Hampson, Photochemical war.

396 Crutzen/Birks, Atmosphere.

a highly politically charged scenario. That was all the more the case as the essay was published at the time of the rearmament debate, which burgeoned following the NATO Double-Track Decision of December 12, 1979, and reached its peak in the mid-1980s. The decision combined the planned stationing of 108 Pershing II missiles and 464 cruise missiles in Europe with the offer to negotiate with the USSR on mutual disarmament of nuclear weapons.[397]

The primary feature of the Nuclear Winter hypothesis was that the approach was based on a computer simulation that fortunately has never been put to the test in reality. The authority of the hypothesis was underpinned by additional model studies whose results were not considered entirely reliable but nevertheless point in a similar direction: a massive use of nuclear weapons would overall result in the consequences outlined by Crutzen and Birks in 1982. Major contributions in this context came from Richard Peter Turco (born 1943) of the University of California, scientist and journalist Carl Sagan (1934 – 1996) and others. The model-based work of 1983, later referred to as the TTAPS Study based on the surnames of the authors, came to the conclusion that a nuclear war would first obscure the Earth's surface, second, temperatures below freezing would be expected over several months, and third, substantial weather modifications would occur on a local level. This scenario was referred to as the "Nuclear Winter".[398] The TTAPS team later drew attention to the combustion potential of woodland and construction timber, crude and refined oil, plastic and

397 The field of history has four interpretations of the NATO Double-Track Decision/ the political aims of NATO in the Cold War. While the approaches involving NATO's desire for a consensus in security policy and détente-policy revisionism can themselves fundamentally be regarded as part of the contemporary debate, the approaches of synthesis of the history of society and of internationalism are interested above all in answering historical questions with various prioritizations. For extensive insight into the debate, see the anthology: Gassert et al., Zweiter Kalter Krieg.

398 Turco et al., Nuclear Winter, 1290.

polymer substances, asphalt surfaces and vegetation as specific central factors.[399]

The work of Crutzen, Birks, Turco, Sagan and others brought with it lasting consequences in international science as well as in public debates and political action. Organizational structures were also created on an international scale, such as the "Environmental consequences of Nuclear War" (ENUWAR), a committee established in 1982 in the scope of the "Scientific Committee on problems of the environment" (SCOPE) of the International Council of Scientific Unions (ICSU), which included more than 300 scientists from around the world.[400]

The snappy term Nuclear Winter resonated well with the public.[401] The American science magazine "Discover" voted Crutzen as Scientist of the Year 1984 for his pioneering work on the Nuclear Winter,[402] which was also picked up by the German media.[403] Crutzen's legacy reveals that many readers subsequently contacted Crutzen directly by post with a wide range of comments and requests relating to the Nuclear Winter. It is possible that this is what finally encouraged him to work journalistically as well to make the Nuclear Winter accessible to a broader audience. A good example is the anthology "Schwarzer Himmel – Auswirkungen eines Atomkriegs auf Klima und globale Umwelt" ("Black Sky—Effects of a Nuclear War on Climate and the Global Environment"),[404] which he published in 1986 together with Jürgen Hahn, who had worked under Junge at the MPIC. In addi-

399 See Turco et al., Climate and Smoke, 169.

400 See Pittrock et al., Environmental Consequences.—Paul Crutzen is one of the editors.

401 Badash talkes about a real shock to the public. See Badash, Nuclear Winter's Tale, 4.

402 See Overbye, Prophet.

403 See for example: the cover of the Jülicher Zeitung, December 29, 1984.—Rhein-Zeitung, No. 300/December 27, 1984.—A large dossier on the nuclear winter was published at the beginning of 1985 in the Zeit: see Die Zeit, No. 3/1985, p. 10.

404 Crutzen/Hahn, Schwarzer Himmel.

tion, Crutzen engaged himself politically. For example, he signed the declaration "Wir warnen vor der strategischen Verteidigungsinitiative" ("We Warn Against the Strategic Defense Initiative"), which was directed against the "Strategic Defense Initiative" (hereafter referred to as the SDI program) introduced into the discussion by the Reagan administration and sent to the Federal Chancellor and several ministries in mid-1985.[405] In October, the initiative was again expanded, when several political and cultural leaders joined in, including well-known personalities such as then Minister President of NRW and later Federal President Johannes Rau (1931 – 2006), journalist, publicist and feminist Alice Schwarzer (born 1942) and left-wing intellectual songwriter Hannes Wader (born 1942).[406] The White House initially used the Nuclear Winter as the justification for the SDI program: if it were possible to face a Soviet nuclear strike in the air, the dust otherwise resulting from the impacts would fail to appear. The technical possibilities, however, would not have allowed the numerous Russian nuclear missiles to be repelled thoroughly enough for this program to be able to prevent the scenario of a Nuclear Winter.[407]

On the level of international politics, an important contribution in the context of nuclear disarmament during the final phase of the Cold War in the 1980s can to a certain extent be attributed to the "Nuclear Winter". In 1986, the Pentagon did not consider the effects of a nuclear exchange to be as high as estimated by the studies from the beginning of the 1980s, but nonetheless it perceived them as sufficiently threatening.[408] Mikhail Gorbachev himself admitted that the theory had a certain influence on his own political stance.[409] In 1988, the

405 Staudinger to Crutzen, August 15, 1985, in: AMPG, III. Abt., ZA 125, No. 4-I.— Appell gegen Waffen im Weltraum (1985), AMPG, III. Abt., ZA 125, No. 4-I.

406 See Starlinger to Crutzen, October 10, 1985, in: AMPG, III. Abt., ZA 125, No. 4-II.

407 See Robock, Policy Implications, 360.

408 See Badash, Nuclear Winter's Tale, 165.

409 See Robock, Nuclear Winter, 425.

agreement on "Intermediate Range Nuclear Forces" (INF) that had been signed by Ronald Reagan and Gorbachev in the previous year came into force, initiating nuclear disarmament in the US and the USSR.[410] In the same year, the UN recognized the Nuclear Winter as a scientifically established theory.[411] The number of potential ready-to-use nuclear weapons around the world dropped dramatically from approximately 70,000 warheads in the 1980s to around 8,500 in the 2000s. At the same time, this supply is still more than enough to cause a Nuclear Winter. Moreover, the number of warheads that could hypothetically be put to use again is estimated at 15,000 for Russia and the United States alone.[412]

3.4.4 The Anthropocene and Responsibility: Is geo-engineering a way out?

The work on the influence of humans on the atmosphere and the global climate is a recurring theme throughout Crutzen's biography: from studies on the influence of aviation via CFCs, the ozone hole, greenhouse gases, (anthropogenic) biomass combustion and the Nuclear Winter scenario, to topics not described in detail here, such as methane emissions from mass animal breeding[413] and the emission of nitrous oxide from land treated with fertilizers.[414] As already explained, in 2000 he finally proposed the term Anthropocene, a designation for a new geological era that is currently gaining increasing appeal in scientific and public discussions.

---

410 See Robock, Policy Implications, 361.

411 See archive material provided online on the website of the UN: <http://www.un.org/en/ga/search/view_doc.asp?symbol=A/43/351>, status: May 23, 2018.—See also: Robock, Policy Implication, 360.

412 See Robock, Nuclear Winter, 419.

413 Amongst others: Schade/Crutzen, Emission; Berges/Crutzen, Estimates.

414 See Crutzen/Ehhalt, Effects.

Crutzen himself clearly connects the concept with a call for action involving solutions that are both political as well as from the non-political sphere. The Montreal Protocol regarding the restriction of CFCs is without doubt an example of a relatively successful response on an international political level. In other areas, such successes have largely failed to appear, for example in the case of restricting CO_2 production. For Crutzen, therefore, the question of the responsibility of humankind remains prominent to this day.[415] In 2006, frustrated by the failure of political instruments, in particular in the case of reducing greenhouse gases,[416] Crutzen, in the search for alternatives that did not involve political decision-making processes, triggered a heated debate. His proposal at the time was to intervene in the Earth's system with scientific methods to prevent a climate-based collapse or in other words: we should consider targeted geoengineering.[417]

In essence, the idea of geoengineering was not new and serious efforts to manipulate weather on a local scale had been made since the late 1940s, certainly advanced by the upcoming Cold War[418] and even promoted in the FRG.[419] Geoengineering, however, reached a new dimension with Crutzen's 2006 proposal to manipulate global climate. His first and subsequently highly controversial idea for such an intervention was to investigate whether the release of sulfur compounds into the atmosphere could significantly increase the Earth's potential to reflect sunlight into the space. Crutzen's suggestion explicitly did not relate to a practical implementation of this or a similar approach, but instead expressed that geoengineering should not be treated as taboo and that it is imperative to discuss any options and to develop

415 Carsten Reinhardt/Gregor Lax: Interview with Paul Crutzen, November 17, 2011.

416 Ibid.

417 Crutzen, Albedo Enhancement.

418 See Bonnheim, History, 893.—See also in particular for the USSR: Oldfield, Climate modification.

419 See Achermann, Eroberung.

and research suitable approaches. Initially, this proposal was met with some very heated reactions, based not only on justified criticism of the feasibility and potentially disastrous consequences of injecting sulfur into the atmosphere, but above all because of the basic idea of actively intervening with nature on a large scale.[420]

In the meantime, a serious scientific debate about geoengineering had been launched and some suggestions for operationalization have been made that recently have even been addressed in policy in the FRG, in particular by the Federal Ministry for the Environment[421]. The proposals range from injecting aerosols into the atmosphere to solar sails in space to redirect a portion of sunlight.[422]

420 Carsten Reinhardt/Gregor Lax: Interview with Paul Crutzen, November 17, 2011.

421 Umweltbundesamt, Geo-Engineering.

422 See Low et al.: Climate Engineering, 3.

FINDINGS AND OUTLOOK

The MPIC's present-day research profile developed, over the course of four decades starting at the end of the 1960s, from innovative atmospheric research that had not played any notable role anywhere in the MPG until the establishment of the Department for Atmospheric Chemistry. Focus was given to increasingly integrative approaches that no longer gave attention to the atmosphere alone, but instead to the atmosphere's mutual exchange processes and interactions with other Earth subsystems. This perspective was reflected in both the epistemic and the academic-organizational development. Considering other subsystems, such as the biosphere and the oceans, made it necessary to incorporate other research areas under the umbrella of atmospheric and Earth System research. At the organizational level, this was reflected at the MPIC in a gradual expansion of Earth-system-oriented departments, starting with biogeochemistry in 1987 and ultimately resulting in an overall institutional alignment that led in the 2000s to the formation of other spin-off branches such as multiphase chemistry.

The foundation for this development was set in 1968 almost by chance, after a dry spell lasting nearly a decade, throughout which the responsible committees at the MPG were not able to acquire a candidate for the director position. One of the causes for this was the long-lasting difficulties in reaching agreement on future topic areas and finding suitable management staff for them. Josef Mattauch had assumed responsibility for the expansion of mass spectrometry at the

MPIC and initially a successor was looked for with a direct connection to the topics of his works.[423] After multiple setbacks, the responsible committees were gradually forced to become more open with regard to the choice of candidate and ultimately also in the possible topic orientation. Out of necessity, the institute tradition had to be postponed through lack of suitable personnel. Innovative alignments only became viable after the closure of the institute was being considered. This process ended with the establishment of atmospheric chemistry in 1968 and also with the strengthening of cosmochemistry, which had been introduced by Friedrich Paneth back in 1953. In retrospect, the latter is an irony of history, as the cosmochemists that remained after Paneth's death had serious difficulties in obtaining support for their research from the MPG throughout the prolonged episodes of the search for a successor. Atmospheric chemistry and cosmochemistry prevailed alongside each other for almost four decades and had a crucial influence on the MPIC. After Günter Lugmaier was conferred emeritus status in 2005, however, cosmochemistry was ultimately discontinued and since 2012 the institute has exclusively dealt with Earth-system questions with an atmospheric science focus across departments.

The new establishment of integrative atmospheric chemistry was the result of a complex constellation and would likely have been inconceivable back in 1960 (chapter 1). The field was at the time a long way off from becoming established for the long term in the atmospheric sciences of the Federal Republic of Germany, which was dominated by classic weather research. The social and political interest in environmental science issues (e. g. acid rain) that had been growing gradually since the second half of the 1960s increased the general attention given to atmospheric sciences. This undoubtedly facilitated the perception and interest in the committees responsible for the appointments and topic selection at the MPG, which had just begun at this

423 See Reinhardt, Massenspektroskopie, 111.

time to, generally and more intensely, address questions relating to the social importance of research. But mundane issues also played a role in the appointment of Junge: The suitable candidate was anchored at Gutenberg University, with which there had been collaboration projects for years and which was located almost at the doorstep of the MPIC. The appointment then took place at record speed relative to the standard appointment procedures at the MPG, after ten years of not being able to appoint a director at the institute.

After Junge's department was opened, integrative atmospheric research gradually began to expand—and at first to the outside. In combination with the DFG-SFB 73, which had been funded since 1970, Junge's department which was increasingly networking with other national and international institutions, became a driver for the expansion of atmospheric sciences at the MPG, but also in the Federal Republic as a whole (chapter 2). During Junge's term, no further departments were initially founded at the MPIC itself that would have further influenced the Earth-system profile of this institute. At a higher level, however, Junge was significantly involved in the greater integration of meteorology at the MPG since the 1970s. These endeavors were reflected at the institutional level in 1975 with the foundation of the Max Planck Institute for Meteorology (MPI-M) in Hamburg. The institute was initially founded with two departments: "Ozeanische Prozesse und Planmodelle" ("Oceanic Processes and Planning Models") under the leadership of Klaus Hasselmann (born 1931) and "Physikalische Prozesse in der Atmosphäre" ("Physical Processes in the Atmosphere") under Hans Georg Hinzpeter (1921 – 1999). The institute is now made up of three departments, all of which have the phrase "Earth System" in the title: The Atmosphere in the Earth System under Björn Stevens (born 1966), The Land in the Earth System (Martin Claußen) and The Ocean in the Earth System (Jochem Marotzke).[424]

424 Cf. the web presence of the MPI-M: <https://www.mpimet.mpg.de/en/institute/organisation/>. Status: May 23, 2018.

In cooperation with the University of Hamburg, the International Max Planck Research School on Earth System Modeling has also been established at the MPI-M. The precise circumstances and background of this institute's foundation and development, in particular with a view to the expansion of the German Climate Computing Center and the further forcing of integrative approaches, are a major component of current research in connection with the GMPG subproject "Earth System Sciences at the MPG". A publication on this topic is currently in preparation.

At the MPIC, however, cosmochemistry initially dominated into the 1980s, as reflected by the departments. Directly after Christian Junge's appointment in 1968, the institute was made up of the departments of Atmospheric Chemistry, Cosmochemistry, Mass Spectroscopy and Isotope Cosmology, along with Nuclear Physics. In 1979 the department for Geochemistry was set up under Albrecht Werner Hofmann, resulting in the institute structure being divided until the mid-1980s into the departments Atmospheric Chemistry (Crutzen), Geochemistry (Hofmann), Cosmochemistry focusing on moon research and planetology (Heinrich Wänke) along with Isotope Cosmology focusing on mass spectroscopy trace analyses and isotope frequencies in meteorites (Friedrich Begemann). The last two departments in particular cultivated lively exchange and cooperation, both because their topic fields partially overlapped and because of the personal relationships between the two heads. In addition to these four departments, there were two additional working groups (nuclear physics and physical chemistry), which were to a certain extent a remnant of the nuclear-physics strand that was anchored at the institute up to 1978.

Paul Crutzen's inauguration as director at the MPIC in 1980 brought about a generational change at the management level of the Department for Atmospheric Chemistry in various respects. Junge had primarily been an empiricist and under him computer simulation in particular had never achieved any great importance. Crutzen, who was an engineer by nature, had always utilized computer programming and computer models since he joined atmospheric research in the late 1950s, and under him this programming, which has been such a central area in atmospheric and Earth-system research since the 1980s, was significantly strengthened.

With the appointment of Meinrat O. Andreae, which was considerably endorsed by Crutzen, and the corresponding founding of the Department for Biogeochemistry, the next decisive step at the structural level of the institute was taken in 1987, which was to expand and influence the Earth-system profile in the long term. The departments of Atmospheric chemistry and Biogeochemistry established from the outset close collaboration at local to international levels, in particular with regard to the examination of substance cycles and of large-scale biomass burning projects (chapter 3).

Another factor that characterized Crutzen's term much more visibly than Junge's was an unerring sense for highly charged social and political topics and focusing on conducting research on the related anthropogenic influences on the atmosphere. While Junge had dealt with the latter considerably in the 1970s, for Crutzen it became a central research program that had already prominently influenced his scientific career before his inauguration at the MPIC. This included in particular studies of CFCs and of the importance of NO_x for the decomposition of ozone, which in turn led to the question as to what dangers the stratospheric ozone layer was exposed to by supersonic aircraft. The burning of biomass, "nuclear winter", finding explanations for the Antarctic hole in the ozone layer, climate change, the role of aerosols and propellant gases (in particular CO_2, CH_4, N_2O, NO_x) and research into their cycles in and between the biosphere, geosphere and atmosphere were the major atmospheric and Earth-system science topic areas covered by Crutzen's—and some of them later also Andreae's—department. Almost all of these topics were important not only in the context of scientific discourse but also in more general social and political discourse; the research sometimes had as a consequence serious political actions, such as in the case of the global reduction of the production and use of CFCs. Scientists need not necessarily be politically active themselves just because their works are of political importance. Crutzen, however, can undoubtedly in retrospect be considered a "political" scientist who very actively participated in these discourses beyond the research itself. This was reflected particularly clearly in his diverse committee and appointment activities and in his own popular science writings. Many areas worked on by Crutzen were the subject of significant media attention even before the publications that were highly popular with the public, as shown by the example of nuclear winter. Crutzen, Molina, and Rowland being awarded the first and so far the only Nobel Prize for atmospheric research in 1995 should be seen not only as an appreciation of this topic area, but also as a highly effective increase in media attention for highly politically charged topic areas from atmospheric research.

A further crucial thrust after the expansion of the MPIC and the establishment of the MPI-M that once again allowed Earth-system sciences to expand significantly at the MPG took place in the 1990s in the course of the rebuilding of East Germany. Following the reunification of Germany, the MPG was confronted with the need to consider the option of setting up new institutes in the new German states. To this end, the members of the three specialist sections were to submit suggestions for topics that should or could help to shape the MPG's future research profile. The submissions included an intersectional proposal from the members of the MPG's CPT section (including Crutzen and MPI-M director Lennart Bengtsson), who had consulted with the members of the biomedical section (BM section). The proposers called for research into biogeochemical cycles to be pushed more strongly than previously at the MPG. In a report to the CPT section, Crutzen explained the central considerations that had led to this proposal. The report stated that recent decades had made it clear *"that the chemical composition of the atmosphere and activities of the biosphere [form] a system with mutual influence"*.

> "On the basis of this finding, efforts should be made to move away from the approach previously taken in the research of the global environment—that of examining the (Earth's systems, atmo-, bio-, geo-) to a great extent separately—and to integrate findings concerning individual processes into a global system."[425]

Pointing out that addressing such questions would give the institute a position that was unique in Europe[426] must have further contributed to action following in 1996/1997. The proposal ultimately led to MPI-BGC being founded in Jena. This institute and the institutional and personnel structures, networks and interconnections at the MPG, that contributed to its founding will also be the subject matter of future works in connection with the GMPG "Earth System" subproject.

The MPIC, however, went on to be restructured again in the mid-2000s. The previously mentioned closure of Cosmochemistry in 2005 was followed at the institute in 2007 by the dissolution of the Department for Geochemistry, when Albrecht W. Hofmann was conferred emeritus status, although it was transferred to Andreae's department rather than being fully closed.[427] This meant that the path was finally clear for completly focusing on Earth-system topic areas, which has continued to this day at the MPIC, even after the closure of Andreae's department in 2016. The institute's goal, according to the description it gives of itself on its website, is to achieve *"an integral scientific understanding of chemical processes in the Earth System from molecular to global scales"*.[428]

---

425 AMPG, II. Abt., Rep. 62, No. 1836, Bl. 22.

426 Cf. ibid.

427 Cf. Kant/Lax, Chronik, 278 f.

428 Web presence of the MPIC: <https://www.mpic.de/ueber-uns.html>. Status: February 7, 2018.

AUTHOR

Gregor Lax's main interests lie in the history of science policy, of organizational, institutional and structural developments in science as well as the genesis, development and power of science policy discourse formations. After completing the MA program "History, Philosophy and Sociology of Sciences" (HPSS) he received his doctorate from the University of Bielefeld in 2014. The focus of his doctoral thesis was on the relationship between basic and applied research in the context of science policy and innovation in the Federal Republic of Germany since 1945. From 2011 to 2014 Lax worked as a research scholar at the Max Planck Institute for Chemistry (MPIC) in Mainz as part of a project on the institute's history. The research focus of his postdoctoral projects between 2014 and 2016, which were also supported by the MPIC, was on the history of the atmospheric sciences in Germany. In 2016–2017 he worked as a science coordinator for the German Council of Science and Humanities (Wissenschaftsrat), where he supervised the evaluation of extensive research infrastructure projects, especially in the field of atmospheric and Earth system sciences, for a national roadmap for the Federal Ministry of Education and Research. In addition, he was concerned with strategies for the promotion of excellence, transfer processes as well as the current debate on a potential erosion of trust in science. In January 2018, Gregor Lax started working as a Research Scholar in the GMPG Research Program on the history of Earth system research in the MPG.

ACKNOWLEDGMENTS

This project was significantly promoted and supported by MPIC staff members. At this point, I would like to give special thanks to Meinrat O. Andreae, Ulrich Pöschl, Paul Crutzen, Susanne Benner, Tracey Andreae, Simone Schweller and Andreas Zimmer. Further, I would like to cordially thank Carsten Reinhardt from Bielefeld University, Rebecca Mertens, David Rengeling and the researchers and the fantastic student assistants of the program concerning the history of the Max Planck Society at the MPI for the History of Science for their constructive discussions and friendly willingness to read. For their willingness to participate for the insightful interviews, I also owe great thanks to Hans-Walter Georgii, Günter Herrmann, Norbert Trautmann, Friedrich Begemann, Heike Tilzer and once again Meinrat O. Andreae and Paul Crutzen.

APPENDIX

Figures

Cover left: Biomass burning (photo by Jack Fishman). Top left: Christian Junge (Archiv der Max-Planck-Gesellschaft, Berlin-Dahlem). Top right: Paul Crutzen (Carsten Costard). Middle: MPI for Chemistry. Bottom middle: Meinrat O. Andreae (Danny Rosenberg). Bottom right: Tracey Andreae (Meinrat O. Andreae).

Fig. 1: AMPG.

Fig. 2: AMPG.

Fig. 3: AMPG.

Fig. 4: L. Schütz, in: Kant/Reinhardt, 100 Jahre, 198.

Fig. 5: H. Elbert, in: Kant/Reinhardt, 100 Jahre, 151.

Fig. 6: Gregor Lax.

Fig. 7: AMPG.

Fig. 8: AMPG.

Fig. 9: Hannes Grobe, Wikimedia Commons: <https://commons.wikimedia.org/wiki/File:Meteor-1964_hg.jpg>, status May 23, 2018 (GFDL, CC-BY 2.5).

Fig. 10: Paul Crutzen and P. Zimmermann, in: Galbally: International Global Atmospheric Chemistry, Front Cover.

Fig. 11: R. A. Kerr, in: Kant/Reinhardt, 100 Jahre, 163.

Fig. 12: Ross Salawitch, NASA, Wikimedia Commons: <https://commons.wikimedia.org/wiki/File:Arctic_stratospheric_cloud.jpg>, status: May 23, 2018.

Unpublished literature

Federal Archive Koblenz

German Research Society

BArch, B227/10803, 010802, 010919, 011037, 011461, 011244, 011460, 012081, 012082, 077088.

Reichsministerium des Inneren

BArch, R1501/ZA VI 0087 A.02/03.

Archive of the Max Planck Society

MPI for Chemistry

Abt. II Rep. 25 (A50), file 13 and file 14.

Senate

AMPG, II. Abt., Rep. 60, No. 50.SP, 51.SP, 53.SP, 55.SP, 59.SP, 62.SP, 85.SP, 89.SP, 90.SP.

Wissenschaftlicher Rat

AMPG, II. Abt., Rep. 62, No. 494, 495, 498, 1771, 1775, 1836.

Institutsbetreuer (General Administration)

AMPG, II. Abt., Rep. 66, No. 840, 841, 842, 843, 844, 848, 849, 850, 851, 854.
AMPG, II. Abt., Rep. 1A IB, Chem 1.6.

Christian E. F. Junge Papers

AMPG, III. Abt., ZA 95, folders 1 – 6, 8.

Paul J. Crutzen Papers

AMPG, III. Abt., ZA 125, No. 4, Teil I – II.
AMPG, III. Abt., ZA 125, No. 9.

Private documentation

Interviews

Lax, Gregor: Telephone-Interview with Hans-Walter Georgii, April 27, 2015 in Bielefeld and Oberursel.
Lax, Gregor: Interview with Heike Tilzer, August 14, 2015 in Konstanz.
Lax, Gregor: Interview with Meinrat O. Andreae, December 2, 2015 in Mainz.
Kant, Horst, Gregor Lax and Anja Heller: Interview with Günter Herrmann und Norbert Trautmann, April 28, 2012 in Mainz.
Reinhardt, Carsten and Gregor Lax: Interview with Paul Crutzen, November 17, 2011 in Mainz.
Reinhardt, Carsten and Gregor Lax: Interview with Friedrich Begemann, January 6, 2012 in Mainz.

E-Mail

Christoph Brühl (MPIC) to Gregor Lax, November 9, 2015.

Published literature

100 Gramm Mond nach Mainz. Interview mit Professor Heinrich Wänke vom Max-Planck-Institut für Chemie in Mainz. *Der Spiegel*, 14. Juli 1969, 29, 105.
Achermann, Dania: Die Eroberung der Atmosphäre. Wetterbeeinflussung in Süddeutschland zur Zeit des Kalten Krieges. *Technikgeschichte* 80 (2013), No. 3, 225 – 239.

Achermann, Dania: Institutionelle Identität im Wandel. Zur Geschichte des Instituts für Physik der Atmosphäre in Oberpfaffenhofen. Bielefeld: Transcript 2016.

Andreae, Meinrat O.: Distribution and spaciation of arsenic in the natural environment. *Deep-Sea Research* 25 (1978), 391 – 402.

Andreae, Meinrat O.: Biomass burning. Its history, use and distribution and its impact on environmental quality and global climate. In: Joel S. Levine (ed.): Global Biomass Burning. Atmospheric, Climatic and Biospheric Implications. Cambridge, MA: MIT Press 1991, 3 – 21.

Andreae, Meinrat O.: Biogeochemische Forschung am Kaiser-Wilhelm-/Max-Planck-Institut für Chemie. In: Kant/Reinhardt, 100 Jahre, 133 – 185.

Andreae, Meinrat O., Artaxo, P., Fischer, S. et al.: Transport of biomass burning smoke to the upper troposphere by deep convection in the equatorial region. *Geophysical Research Letters* 28 (2001), 951 – 954.

Andreae, Meinrat O.; Marotzke, Jochem; Heimann, Martin: Partnerschaft Erdsystemforschung. Jena, Mainz and Hamburg 2006.

Badash, Lawrence: A Nuclear Winter's Tale. Science and Politics in the 1980s. Cambridge, MA: MIT Press 2009.

Barrie, L. A.; Bottenheimer, J. W.; Schnell, R. C. et al.: Ozone destruction and photochemical reactions at polar sunrise in the lower Arctic atmosphere. *Nature* 334 (1988), 138 – 141.

Beck, Silke: Moving beyond the linear model of expertise? IPCC and the test of adaption. *Regional Environmental Change* 11 (2011), 297 – 306.

Begemann, Friedrich; Libby, Willard F.: Continental water balance, groundwater inventory and storage times, surface ocean mixing rates, and world wide water circulation patterns from cosmic-ray and bomb tritium. *Geochimica et Cosmochimica Acta* 12 (1957), 227 – 296.

Berg, W. W.; Crutzen, Paul J.; Grahek, F. E. et al.: First measurements of total chlorine and bromine in the lower stratosphere. *Geophysical Research Letters* 7 (1980), 937 – 940.

Berges, M. G. M.; Crutzen, Paul J.: Estimates of global N_2O emissions from cattle, pig and chicken manure, including a discussion of CH_4 emissions. *Journal Atmospheric Chemistry* 24 (1996), 241 – 269.

Blankenship, James R.; Crutzen, Paul J.: A photochemical model for the space-time variations of the oxygen allotropes in the 20 to 100 km layer. *Tellus* 2/18 (1966), 160 – 175.
Bolin, Bert: Atmospheric Chemistry and broad geophysical relationships. *Proceedings of the National Academy of Sciences of the United States of America* 12/45 (1959), 1663 – 1672.
Bolin, Bert; Crutzen, Paul J.; Vitousek, P. M. et al.: Interactions of biogeochemical cycles. In: Bert Bolin (ed.): The Major Biogeochemical Cycles and Their Interactions. Chichester: Wiley 1983, 1 – 40.
Bonnheim, Norah B.: History of climate engineering. *WIRE's Climate Change* 1 (2010), 891 – 897.
Böschen, Stefan: Risikogenese. Prozesse gesellschaftlicher Gefahrenwahrnehmung: FCKW, Dioxin, DDT und Ökologische Chemie. Opladen: Leske und Budrich 2000.
Brüggemann, Julia: Die Ozonschicht als Verhandlungsmasse. Die deutsche Chemieindustrie in der Diskussion um das FCKW-Verbot 1974 bis 1991. *Zeitschrift für Unternehmensgeschichte* 2/60 (2015), 168 – 193.
Brühl, Christoph; Crutzen, Paul J.: Scenarios of possible changes in atmospheric temperatures and ozone concentrations due to man's activities as estimated with a one-dimensional coupled photochemical climate model. *Climate Dynamics* 2 (1988), 173 – 203.
Brühl, Christoph; Crutzen, Paul J.: Ozone and climate changes in the light of the Montreal Protocol. A model study. *Ambio* 19 (1990), 293 – 301.
Brühl, Christoph; Crutzen, Paul J.: The atmospheric chemical effects of aircraft operations. In: Ulrich Schumann (ed.): Air Traffic and the Environment—Background, Tendencies and Potential Global Atmospheric Effects. Proceedings of a DLR International Colloquium, Bonn, Germany, November 15 – 16. Berlin: Springer 1990, 96 – 106.
Brühl, Christoph; Peter, Th.; Crutzen, Paul J.: Increase in the PSC-formation probability caused by high-flying aircraft. *Geophysical Research Letters* 18 (1991), 1465 – 1468.
Chapman, Sydney: A Theory of upper-atmospheric ozone. *Memoirs of the Royal Meteorological Society* 3/26 (1929), 103 – 125.

Chapman, Sydney: On ozone and atomic oxygen in the upper atmosphere. *London, Edinburgh and Dublin Philosophical Magazine and Journal of Science* 10/64 (1930), 369 – 383.

Conway, Erik; Oreskes, Naomi: Merchants of Doubt. How a Handful of Scientists Obscured the Truth on Issues from Tobacco Smoke to Global Warming. New York, NY: Bloomsbury Press 2010.

Crutzen, Paul J.: Determination of parameters appearing in the "dry" and the "wet" photochemical theories for ozone in the stratosphere. *Tellus* 3/21 (1969), 368 – 388.

Crutzen, Paul J.: The influence of nitrogen oxides on the atmospheric ozone content. *Quarterly Journal of the Royal Meteorological Society* 96 (1970), 320 – 325.

Crutzen, Paul J.: SST's—A Threat to the Earth's Ozone Shield. *Ambio* 2/1 (1972), 41 – 51.

Crutzen, Paul J.: The photochemistry of the stratosphere with special attention given to the effects of NO_x emitted by supersonic aircraft. US Department of Transportation, First Conference on CIAP 1972, 880 – 888.

Crutzen, Paul J.: Estimates of possible future ozone reductions from continued use of fluorochloromethanes (CF_2Cl_2, $CFCl_3$). *Geophysical Research Letters* 1 (1974), 205 – 208.

Crutzen, Paul J.: Estimates of Possible Variation in Total Ozone Due to Natural Causes and Human Activities. *Ambio* 6/3 (1974), 201 – 210.

Crutzen, Paul J.: The possible importance of CSO for the sulfate layer of the stratosphere. *Geophysical Research Letters* 3 (1976), 73 – 76.

Crutzen, Paul J.: Geology of Mankind. *Nature* 1/415 (2002), 23.

Crutzen, Paul J.: Albedo Enhancement by Stratospheric Sulfur Injections. A Contribution to Resolve a Policy Dilemma? *Climatic Change* 77 (2006), 211 – 219.

Crutzen, Paul J.; Andreae, Meinrat O.: Atmospheric Chemistry. In: T. F. Malone and J. G. Roederer (ed.): *Global Change*. Cambridge: Cambridge University Press 1985, 75 – 113.

Crutzen, Paul J.; Andreae, Meinrat O.: Biomass Burning in the Tropics. Impact on Atmospheric Chemistry and Biogeochemical Cycles. *Science* 4988/250 (1990), 1669 – 1678.

Crutzen, Paul J.; Arnold, F.: Nitric acid cloud formation in the cold Antarctic stratosphere. A major cause for the springtime "ozone hole". *Nature* 324 (1986), 651 – 655.

Crutzen, Paul; Birks, John: The Atmosphere after a Nuclear War: Twilight at Noon. *Ambio* 11 (1982), 114 – 125.

Crutzen, Paul J., Brühl, Christoph; Schmailzl, U.; Arnold, F.: Nitric acid haze formation in the lower stratosphere: a major contribution factor to the development of the Antarctic "ozone hole". In: M. P. McCormick and P. V. Hobbs (ed.): *Aerosols and Climate*. Hampton, VA: A Deepak 1988, 287 – 304.

Crutzen, Paul J.; Ehhalt, Dieter Hans: Effects of nitrogen fertilizers and combustion on the stratospheric ozone layer. *Ambio* 1 – 3/6 (1977), 112 – 117.

Crutzen, Paul J.; Hahn, Jürgen (ed.): Schwarzer Himmel – Auswirkungen eines Atomkrieges auf Klima und globale Umwelt. Frankfurt am Main: Fischer 1986.

Crutzen, Paul J.; Heidt, Leroy E.; Krasnec, Joseph P. et al.: Biomass burning as a source of atmospheric gases CO, H_2, N_2O, NO, CH_3CL and COS. *Nature* 282 (1979), 253 – 256.

Crutzen, Paul J.; Lax, Gregor; Reinhardt, Carsten: Paul Crutzen on the Ozone Hole, Nitrogen Oxides, and the Nobel Prize. *Angewandte Chemie International Edition* 52 (2013) 1, 48 – 50.

Crutzen, Paul J.; Müller, R.; Brühl, Christoph; Peter, Th.: On the potential importance of the gas phase reaction $CH_3O_2 + ClO \rightarrow ClOO + CH_3O$ and the heterogeneous reaction $HOCl + HCl \rightarrow H_2O + Cl_2$ in "ozone hole". *Geophysical Research Letters* 19 (1992), 1113 – 1116.

Crutzen, Paul J.; Stoermer, Eugene F.: The "Anthropocene". *Global Change Newsletter* 41 (2000), 17 – 18.

Crutzen, Paul J.; Whelpdale, Douglas M.; Kley, Dieter; Barrie, Leonard A.: The cycling of sulfur and nitrogen in the remote atmosphere. In:

James N. Galloway: The Biogeochemical Cycling of Sulfur and Nitrogen in the Remote Atmosphere. Dordrecht: Reidel 1985, 203 – 212.
Dalby, Simon; Lehman, Jessi; Nelson, Sara et al.: After the Anthropocene. Politics and geographic inquiry for a new epoch. *Progress in Human Geography* 38/3 (2014), 439 – 456.
Deckstein, Dinah: Think small. *Der Spiegel*, 7. August 2000, 32, 82 f.
Deutsches Hydrografisches Institut Hamburg (ed.): Forschungsschiff Meteor 1964 – 1985. Hamburg: Deutsches Hydrografisches Institut 1985.
DFG (ed.): Jahresbericht der DFG, Bd. 2, Bad Godesberg 1970.
DFG (ed.): Jahresbericht der DFG, Bd. 2, Bad Godesberg 1976.
DFG (ed.): Jahresbericht der DFG, Bd. 2, Bad Godesberg 1979.
DFG (ed.): Jahresbericht der DFG, Bd. 2, Bad Godesberg 1980.
DFG (ed.): Jahresbericht der DFG, Bd. 2, Bad Godesberg 1982.
DFG (ed.): Jahresbericht der DFG, Bd. 2, Bad Godesberg 1985.
DFG (ed.): Jahresbericht der DFG, Bd. 2, Bad Godesberg 1986.
Edwards, Paul N.: History of Climate Modeling. Wiley Interdisciplinary Reviews: Climate Change 2 (2011), 128 – 139.
Engels, Jens Ivo: Geschichte und Heimat. Der Widerstand gegen das Kernkraftwerk Wyhl. In: Kerstin Kretschmer (ed.): Wahrnehmung, Bewusstsein, Identifikation. Umweltprobleme und Umweltschutz als Triebfedern regionaler Entwicklung. Freiberg: Technische Universität Bergakademie 2003, 103 – 130.
Farman, Joe C.; Gardiner, B. G.; Shanklin, J. D.: Large Losses of Total Ozone in Antarctica Reveal Seasonam ClO_x/NO_x Interaction. *Nature* 315 (1985), 207 – 210.
Fischer, H., Waibel, A. E.; Welling, M. et al.: Observations of high concentration of total reactive nitrogen (NO_y) and nitric acid (HNO_3) in the lower Arctic stratosphere during the Stratosphere-Troposphere Experiment by Aircraft Measurements (STREAM) II campaign in February 1995. *Journal of Geophysical Research* 102 (1997), 23559 – 23571.
Feck, Thomas: Wasserstoff-Emissionen und ihre Auswirkungen auf den arktischen Ozonverlust. Risikoanalyse einer globalen Wasserstoffwirtschaft. Jülich: Forschungszentrum Jülich 2009.

Galbally, Ian E.: The International Global Atmospheric Chemistry (IGAC) Programme. A Core Project of the International Geosphere-Biosphere Programme. Stockholm 1989.

Gassert, Phillipp; Geiger, Tim; Wentker, Hermann (ed.): Zweiter Kalter Krieg und Friedensbewegung: Der NATO-Doppelbeschluss in deutsch-deutscher und internationaler Perspektive. München: Oldenbourg 2011.

Gidel, Louis T.; Crutzen, Paul J.; Fishman, Jack: A two-dimensional photochemical model of the atmosphere. 1: Chlorocarbon emissions and their effect on stratospheric ozone. *Journal of Geophysical Research* 88 (1983), 6622 – 6640.

Gordon, Glen E.: Reviews on Atmospheric Chemistry (Barbara J.Finlayson-Pitts and James N. Pitts) and Atmospheric Chemistry and Physics of Air Pollution (John S. Seinfeld). *Science* 235/4793 (1987), 1263 f.

Gramelsberger, Gabriele: Conceiving process in atmospheric models—General equations, subscale parameterizations, and "superparameterizations." *Studies in History and Philosophy of Modern Physics* 41 (2010), 233 – 241.

Gramelsberger, Gabriele; Feichter, Johann (ed.): Climate Change and Policy. The Calculability of Climate Change and the Challenge of Uncertainty. Berlin: Springer 2011.

Grooß, J. U.; Peter, Th.; Brühl, C.; Crutzen, P. J.: The Influence of high flying aircraft on polar heterogeneous chemistry. In: U. Schuhmann, D. Wurzel (ed.): Proceedings of an international scientific Colloquium on Impact of Emissions from Aircraft and Spacecraft upon the Atmosphere. Köln 1994, 229 – 234.

Grundmann, Reiner: Transnational Environmental Policy. Reconstructing ozone. London: Routledge 2001.

Hahn, Jürgen: N_2O Measurements in the Northeast Atlantic Ocean. *Meteor* Reihe A (1975), 1 – 14.

Hall, R. Cargill: A History of the Military Polar orbiting Meteorological Satellite Program. Chantilly: National Reconnaissance Office 2001.

Hampson, John: Chemiliuminescent emissions observed in the stratosphere and mesosphere. In: Les problèmes météorologiques de la stratosphere et de la mésosphère. Paris: Presses Universitaires de France 1966, 393 – 440.

Hampson, John: Photochemical war on the atmosphere. *Nature* 250 (1974), 189 – 191.

Hart, David M.; Victor, David G.: Scientific Elites and the Making of US Politics for Climate Change Research, 1957 – 1974. *Social Studies of Science* 23 (1993), 643 – 680.

Heymann, Matthias: Lumping, testing, tuning: The invention of an artificial chemistry in atmospheric transport modeling. *Studies in History and Philosophy of Modern Physics* 41 (2010), 218 – 232.

Heymann, Matthias: Understanding and misunderstanding computer simulation: The case of atmospheric and climate science—An introduction. *Studies in History and Philosophy of Modern Physics* 41 (2010), 193 – 200.

Hoffmann, Dieter; Schmidt-Rohr, Ulrich (ed.): Wolfgang Gentner – Festschrift zum 100. Geburtstag. Berlin: Springer 2006.

Hoffmann, Dieter; Kolboske, Birgit; Renn, Jürgen (ed.): "Dem Anwenden muss das Erkennen vorausgehen". Auf dem Weg zu einer Geschichte der Kaiser-Wilhelm/Max-Planck-Gesellschaft. Berlin: epubli 2014, 133 – 191.

Hornung, Helmut: Den Mond in der Nase. *Max Planck Forschung* 4 (2010), 96 – 97.

Houghton, John T.; Filho, Luiz Meira G.; Callander, B. A. et al. (ed.): Climate Change 1995. The Science of Climate Change. Contribution of Working Group I to the Second Assessment Report of the Intergouvernmental Panel on Climate Change. Cambridge University Press 1996.

Jaenicke, Ruprecht: Die Erfindung der Luftchemie – Christian Junge. In: Kant/Reinhardt, 100 Jahre, 187 – 202.

Jardine, Kolby; Yañez-Serrano, A. M.; Williams, J. et al.: Dimethyl sulfide in the Amazon rain forest. *Biogeochemical Cycles* 29 (2015), 19 – 32.

Jochum, Klaus Peter: Drei Jahrzehnte Funkenmassenspektrometrie (SSMS) im Bereich Geo- und Kosmochemie am Max-Planck-Institut für Chemie Mainz. In: Klaus Peter Jochum, Brigitte Stoll, and Michael Seufert (ed.): 20 Jahre Arbeitstagung "Festkörpermassenspektrometrie" (1977 – 1997). Mainz: Max-Planck-Institut für Chemie, 2nd Edition 1998, 1 – 58.

Johnson, Jeffrey A.: The Kaiser's Chemists. Science and Modernization in Imperial Germany. Chapel Hill, NC: University of North Carolina Press 1990.

Jülicher Zeitung (ed.): Titelblatt. *Jülicher Zeitung* 29.12.1984.
Junge, Christian: Air Chemistry and Radioactivity. New York: Academic Press 1963.
Junge, Christian: Chemical Composition of Precipitation. *Air Chemistry Radioactivity* 4 (1963), 289–310.
Junge, Christian E.: Chimičeskij sostav i radioaktivnost' atmosfery. Moskau: Izd. Mir 1965.
Junge, Christian: Survey about our present Knowledge of Atmospheric Aerosols with Respect to their Role in Cloud Physics. *Bulletin of the American Meteorological Society* 5P2/49 (1968), 592 ff.
Junge, Christian: Der Stoffkreislauf der Atmosphäre. Probleme und neuere Ergebnisse der luftchemischen Forschung. In: Jahrbuch der Max-Planck-Gesellschaft zur Förderung der Wissenschaften e. V. 1971. Göttingen: Hubert & Co. 1971, 149–181.
Junge, Christian: Atmospheric Aerosols and Cloud Formation – Status of Knowledge and Problems. *Kolloid-Zeitschrift und Zeitschrift für Polymere* 250/7 (1972), 638 f.
Junge, Christian: Cycle of atmospheric Gases, Natural and Man Made. *Quarterly Journal of the Royal Meterological Society* 418/98 (1972), 711–729.
Junge, Christian: Our Knowledge of Physico-Chemistry of Aerosols in undisturbed Marine Environment. *Journal of Geophysical Research* 27/77 (1972), 5183–5200.
Junge, Christian: Die Entstehung der Erdatmosphäre und ihre Beeinflussung durch den Menschen. In: Generalverwaltung der Max-Planck-Gesellschaft zur Förderung der Wissenschaften e. V. (ed.): Max-Planck-Gesellschaft Jahrbuch 1975. Göttingen: Vandenhoeck & Ruprecht 1975, 36–48.
Junge, Christian: Stable Isotope Fractionation in Geochemical and Environmental Cycles. In: Werner Stumm: Global Chemical Cycles and their Alterations by Man. Berlin: Abakon 1977, 33–44.
Junge, Christian E.; Chagnon, Charles W.; Manson, James E.: Stratospheric aerosols. *Journal of Meteorology* 18 (1961), 81–108.

Junge, Christian; Hahn, Jürgen: N_2O Measurements in North Atlantic. *Journal of Geophysical Research* 33/76 (1971), 8143 – 8146.

Junge, Christian E.; Manson, James E.: Stratospheric Aerosol Studies. *Journal of Geophysical Research* 66/7 (1961), 2163 – 2182.

Junge, Christian; McLaren; Eugene: Relationship of Cloud Nuclei Spectra to Aerosol size Distribution and Composition. *Journal of the atmospheric sciences* 3/28 (1971), 382 – 390.

Junge, Christian; Seiler, Wolfgang; Schmidt, Ulrich et al.: Kohlenmonoxid- und Wasserstoffproduktion mariner Mikroorganismen im Nährmedium mit synthetischem Seewasser. *Die Naturwissenschaften* 11/59 (1972), 514 f.

Junge, Christian; Werby, R. T.: The Concentration of Chloride, Sodium, Potassium, Calcium and Sulphate in Rain Water over the United States. *Journal of Meteorology* 15 (1958), 417 – 425.

Kant, Horst et al.: Die Wissenschaftlichen Mitglieder des Kaiser-Wilhelm-/ Max-Planck-Instituts für Chemie (Kurzbiographien). In: Kant/Reinhardt, 100 Jahre, 307 – 367.

Kant, Horst; Lax, Gregor: Chronik des Kaiser-Wilhelm-/Max-Planck-Instituts für Chemie. In: Kant/Reinhardt, 100 Jahre, 261 – 277.

Kant, Horst; Reinhardt, Carsten (ed.): 100 Jahre Kaiser-Wilhelm-/Max-Planck-Institut für Chemie (Otto-Hahn-Institut). Facetten seiner Geschichte. Berlin: Archiv der Max-Planck-Gesellschaft 2012.

Kazemi, Marion: Nobelpreisträger in der Kaiser-Wilhelm-/Max-Planck-Gesellschaft zur Förderung der Wissenschaften. Berlin: Archiv der Max-Planck-Gesellschaft, 2nd Edition 2006.

Klee, Ernst: Das Personenlexikon zum Dritten Reich. Wer war was vor und nach 1945. Frankfurt am Main: Fischer 2005.

Krafft, Fritz: Im Schatten der Sensation. Leben und Wirken von Fritz Straßmann. Weinheim: Verlag Chemie 1981.

Küppers, Günter, Peter Lundgreen, and Peter Weingart: Umweltforschung – die gesteuerte Wissenschaft? Eine empirische Studie zum Verhältnis von Wissenschaftsentwicklung und Wissenschaftspolitik. Frankfurt am Main: Suhrkamp 1978.

Kürschners Deutscher Gelehrtenkalender. 22. Ausgabe 2009. Berlin: De Gruyter 2009.

Kwa, Chunglin: Local Ecologies and Global Science: Discourses and Strategies of the International Geosphere-Biosphere Programme. *Social Studies of Science* 35 (2005), 923 – 950.

Lacaux, J. P.; Delmas, R.; Jambert, C.; Kuhlbusch, T. A. J.: NO_x emissions from African savanna fires. *Journal of Geophysical Research* 101 (1996), 23/D19, 585 – 595.

Laitko, Hubert: Das Harnack-Prinzip als institutionelles Markenzeichen. Faktisches und Symbolisches. In: Hoffmann/Kolboske/Renn, Anwenden, 133 – 191.

Lässing, Volker: Den Teufel holt keiner! Otto Hahn und das Kaiser-Wilhelm-Institut für Chemie in Tailfingen. Albstadt: CM-Verlag 2010.

Landsberg, Helmut and L. Machta: Anthropogenic Pollution of the Atmosphere: Whereto? *Ambio* 3/4 (1974), 146 – 150.

Lax, Gregor: Das "Lineare Modell der Innovation" in Westdeutschland. Eine Geschichte der Hierarchiebildung zwischen Grundlagen- und Anwendungsforschung nach 1945. Baden-Baden: Nomos 2015.

Lax, Gregor: Zum Aufbau der Atmosphärenwissenschaften in der BRD seit 1968. *NTM* 24 (2016) 1, 81 – 107.

Leuschner, Anna: Die Glaubwürdigkeit der Wissenschaft. Eine wissenschafts- und erkenntnistheoretische Analyse am Beispiel der Klimaforschung. Bielefeld: Transcript 2012.

Lindner, Stephan: Hoechst. Ein I.G. Farben-Werk im Dritten Reich. München: Beck 2005.

Lovelock, James: Gaia as seen through the atmosphere. *Atmospheric Environment* 8/6 (1972), 579 f.

Lovelock, James: Gaia. A new Look at Life on Earth. Oxford University Press 1979.

Low, Sean; Schäfer, Stefan; Maas, Achim: Climate Engineering. *IASS Fact Sheet* 1 (2013), 1 – 5.

Lüst, Reimar: Der Antriebsmotor der Max-Planck-Gesellschaft: Das Harnack-Prinzip und die Wissenschaftlichen Mitarbeiter. In: Hoffmann/Kolboske/Renn, Anwenden, 119 – 132.

Lustig, Harry: The Mössbauer Effect. *American Journal of Physics* 29 (1961) 1, 1 – 18.
Malm, Andreas; Hornburg, Alf: The geology of mankind? A critique of the Anthropocene narrative. *The Anthropocene Review* 1 (2014), 1 – 8.
Malone, Thomas: Helmut E. Landsberg. Towards AD 2000. In: Ferdinand Baer, Norman Canfield und J. Murray Mitchell: Climate in Human Perspective. A tribute to Helmut E. Landsberg. Dordrecht: Springer Science and Business Media 1991, 21 – 31.
Müller, Edda: Innenwelt der Umweltpolitik – Zu Geburt und Aufstieg eines Politikbereichs. In: Patrick Masius, Ole Sparenberg and Jana Sprenger (ed.): Umweltgeschichte und Umweltzukunft. Zur gesellschaftlichen Relevanz einer jungen Disziplin. Göttingen: Universitätsverlag Göttingen 2009, 69 – 86.
Max Planck Institute for Aeronomie (ed.): Jahresbericht des MPI für Aeronomie. Göttingen 1976.
Oldfield, Jonathan D.: Climate modification and climate change debates among Soviet physical geographers, 1940s – 1960s. *WIRE's Climate Change* 4 (2013), 513 – 524.
Overbye, Dennis: Prophet of the cold and dark. *Discover* 6 (1985), 24 – 32.
Palme, Herbert: Heinrich Wänke und die Erforschung des Mondes und der terrestrischen Planeten. In: Kant/Reinhardt, 100 Jahre, 203 – 239.
Palsson, Gisli; Szerszynski, Bronislaw; Sörlin, Sverker et al.: Reconceptualizing the "Anthropos" in the Anthropocene. Integrating the social sciences and humanities in global environmental change research. *Environmental Science & Policy* 28 (2013), 3 – 13.
Peter, Thomas; Brühl, Christoph; Crutzen, Paul J.: Increase in the PSC-formation probability caused by high-flying aircraft. *Geophysical Research Letters* 18 (1991), 1465 – 1468.
Pittrock, A. B.; Ackerman, Thomas P.; Crutzen, Paul J. et al. (ed.): Environmental Consequences of Nuclear War. New York: Wiley 1986.
Pott, Hermann; Deckstein, Dinah; Jaeger, Ulrich et al.: Richtung Zukunft und zurück. *Der Spiegel*, 31. Juli 2000, 31, 112 – 126.

Rasool, S. I.; Schneider, S. H.: Atmospheric Carbon Dioxide and Aerosols. Effects of Large Increases on Global Climate. *Science* 173 (1971), 138 – 141.

Ravishankara, A. R.; Daniel, J. S.; Portmann, R. W.: Nitrous Oxide (N_2O): The dominant ozone-depleting substance emitted in the 21st Century. *Science* 5949/326 (2009), 123 – 125.

Reinhardt, Carsten: Massenspektroskopie als methodische Klammer des Instituts, 1939 – 1978. In: Kant/Reinhardt, 100 Jahre, 99 – 131.

Revelle, Roger and Hans E. Suess: Carbon Dioxide Exchange between Atmosphere and Ocean and the Question of an Increase of Atmospheric CO_2 during the Past Decades. *Tellus* 1 (1957), 18 – 27.

Robock, Alan: Policy Implications of Nuclear Winter and Ideas for Solutions. *Ambio* 7/18 (1989), 360 – 366.

Robock, Alan: Nuclear Winter. *WIREs Climate Change* 1 (2010), 418 – 427.

Rowland, Frank Sherwood; Molina, Mario: Stratospheric sink for chlorofluoromethanes. Chlorine atom-catalysed destruction of ozone. *Nature* 249/5460 (1974), 810-812.

Rubin, Mordecai B.: The History of Ozone. The Schönbein Period, 1839 – 1868. *Bulletin for the History of Chemistry* 26/1 (2001), 40 – 56.

Rückführung. *Der Spiegel*, 7. August 1963, 32, 12.

Sawyer, J. S: Man-made Carbon Dioxide and the "Greenhouse" Effect. Nature 239 (1972), 23 – 26.

Schaaf, Michael: *Der Physikochemiker Paul Harteck (1902 – 1985)*. Dissertation, Universität Stuttgart 1999.

Schade, Gunnar W.; Crutzen, Paul J.: Emission of aliphatic amines from animal husbandry and their reactions. Potential source of N_2O and HCN. *Journal Atmospheric Chemistry* 22 (1995), 319 – 346.

Schmidt, Ulrich; Seiler, Wolfgang: A new Method for Recording Molecular Hydrogen in Atmospheric Air. *Journal of Geophysical Research* 9/75 (1970), 1713 – 1716.

Schützenmeister, Falk: Zwischen Problemorientierung und Disziplin. Ein koevolutionäres Modell der Wissenschaftsentwicklung. Bielefeld: Transcript 2008.

Schützenmeister, Falk: Offene Großforschung in der atmosphärischen Chemie? Befunde einer empirischen Studie. In: Jost Halfmann and Falk Schützenmeister (ed.). Organisationen der Forschung. Der Fall der Atmosphärenwissenschaft. Wiesbaden: Verlag für Sozialwissenschaften 2009, 171 – 208.

Schwelin, Michael: Nuklearer Winter: Leise rieselt der Schnee. *Die Zeit* 1985, 3, 10.

Seiler, Wolfgang: Der Kreislauf von CO, H_2, N_2O und CH_4. *Promet* 2/5 (1975), 12 – 15.

Seiler, Wolfgang: The Influence of the Biosphere on the Atmospheric Carbon Monoxide and Hydrogen Cycles. In: Wolfgang Krumbein (ed.): *Environmental Biogeochemistry and Geomicrobiology*. Ann Arbor, MI: Ann Arbor Science Publishers 1978, 773 – 810.

Seiler, Wolfgang; Crutzen; Paul J.: Estimates of gross and net fluxes of carbon between the biosphere and the atmosphere from biomass burning. *Climatic Change* 2 (1980), 207 – 247.

Seiler, Wolfgang; Giehl, Helmut: Influence of Plants on Atmospheric Carbon-Monoxide. *Geophysical Research Letters* 7/4 (1977), 329 – 332.

Simpson, H. J.: Man and the Global Nitrogen Cycle Group Report. In: Werner Stumm: Global Chemical Cycles and their Alterations by Man. Berlin: Abakon 1977, 253 – 274.

Singer, S. Fred.: Will the World come to a Horrible End? *Science* 3954/170 (1970), 125.

Stolarski, Richard S., Baughcum, Steven L.; Brune, William H. et al.: The 1995 Scientific Assessment of the Atmospheric Effects of Stratospheric Aircraft. NASA Reference Publication 1381, 1995.

Taba, H.: Bulletin Interview with Helmut E. Landsberg. In: Ferdinand Baer, Norman Canfield, and J. Murray Mitchell: Climate in Human Perspective. A tribute to Helmut E. Landsberg. Dordrecht: Springer Science and Business Media 1991, 97 – 110.

Trischler, Helmuth: The Anthropocene. *NTM* 24/3 (2016), 309 – 335.

Trischler, Helmuth and Rüdiger vom Bruch: Forschung für den Markt. Geschichte der Fraunhofer-Gesellschaft. München: Beck 1999.

Turco, Richard P.; Toon, Owen B.; Ackerman, Thomas P. et al.: Nuclear Winter: Global Consequences of Multiple Nuclear Explosions. *Science* 4630/222 (1983), 1283 – 1292.

Turco, Richard P.; Toon, Owen B.; Ackerman, Thomas P. et al.: Climate and Smoke: An Appraisal of Nuclear Winter. *Science* 247/4939 (1990), 166 – 176.

Uhrqvist, Ola; Lövbrand, Eva: Rendering global change problematic: the constitutive effects of Earth System Research in the IGBP and the IHDP. *Environmental Politics* 23 (2014) 2, 339 – 356.

Uhrqvist, Ola; Linnér, Björn-Ola: Narratives of the Past for Future Earth: The historiography of global environmental change research. *The Anthropocene Review* 2 (2015) 2, 159 – 173.

Umweltbundesamt (ed.): Geo-Engineering. Wirksamer Klimaschutz oder Größenwahn? Methoden – rechtliche Rahmenbedingungen – umweltpolitische Forderungen. Berlin 2011.

Volkert, Hans; Achermann, Dania: Roots, Foundation, and Achievements of the "Institut für Physik der Atmosphäre." In: Ulrich Schumann (ed.): Atmospheric Physics. Background—Methods—Trends. Berlin/ Heidelberg: Springer 2012, 843 – 860.

Wänke, Heinrich; Arnold, James R.: Hans E. Suess 1909 – 1993. *Biographical Memoirs* 87 (2005), 4.

Warneck, Peter: Zur Geschichte der Luftchemie in Deutschland. *Mitteilungen der Fachgruppe Umweltchemie und Ökotoxikologie* 2/9 (2003), 5 – 11.

Watson, Andrew J. and James Lovelock: Biological homeostasis of the global environment. The parable of Daisyworld. *Tellus* 35 B (1983), 284 – 289.

Wege, Klaus: Die Entwicklung der meteorologischen Dienste in Deutschland. Offenbach am Main: Selbstverlag des Deutschen Wetterdienstes 2002.

Weiss, Burghard: Das Beschleunigerlaboratorium am KWI/MPI für Chemie: Kontinuität in deutscher Großforschung. In: Christoph Meinel und Peter Voswinckel (ed.): Medizin, Naturwissenschaft, Technik und Nationalsozialismus. Kontinuitäten und Diskontinuitäten. Stuttgart: GNT-Verlag 1994, 111 – 119.

Weiss, Burghard: The "Minerva" Project. The Accelerator Laboratory at the Kaiser Wilhelm Instituts/Max Planck Institute for Chemistry: Continuity in fundamental research. In: Monika Renneberg und Mark Walker (ed.): Scientists, Engineers and National Socialism. Cambridge: Cambridge University Press 1994, 271 – 290.

Wolf, Christa: Verzeichnis der Hochschullehrer der TH Darmstadt. Darmstadt: Historischer Verein für Hessen 1977.

Zimmermann, Patrick; Chatfield, Robert; Fishman, Jack; Crutzen, Paul J.; Hanst, Phillip L.: Estimates on the production of CO and H_2 from the oxidation of hydrocarbon emissions from vegetation. *Geophysical Research Letters* 5 (1978), 679 – 682.

Internet resources

Anthropocene Working Group of the Subcommission on Qarternary Stratigraphy (International Commission on Stratigraphy) (ed.): Newsletter No. 1 (2009). <http://quaternary.stratigraphy.org/workinggroups/anthropo/Anthropnewsl1.doc>. Status: May 23, 2018.

Bojanowski, Axel; Schwägerl, Christian: Debatte um neues Erdzeitalter: Was vom Menschen übrig bleibt. *Spiegel Online*, 4.7.2011. <http://www.spiegel.de/wissenschaft/natur/a-769581.html>. Status: May 23, 2018.

Bojanowski, Axel: Debatte über Anthropozän: Forscher präsentieren Beweise für neues Menschenzeitalter. *Spiegel Online*, 25.8.2014. <http://www.spiegel.de/wissenschaft/natur/a-987349.html>. Status: May 23, 2018.

Dahan, Amy: Historic Overview of Climate Framing. HAL Workingpapers (2013). <https://halshs.archives-ouvertes.fr/halshs-00855311/document>. Status: June 6, 2018.

DFG-Homepage: Beschreibung der Senatskommission Erdsystemforschung. <http://www.dfg.de/dfg_profil/gremien/senat/erdsystemforschung/>. Status: May 23, 2018.

Die Feuer verändern sich im Minutentakt. *Zeit Online* 23.10.2017. <https://www.zeit.de/gesellschaft/zeitgeschehen/2017-10/kalifornien-wald-braende-san-francisco-tote-suche-vermisste>. Status: October 10, 2018.

Ellis, Erle: Anthropocene. <https://editors.eol.org/eoearth/wiki/Anthropocene>. Status: October 10, 2018.

Harkewicz, Laura: Oral History of Gustaf Olof Svante Arrhenius. 11.4.2006. <http://libraries.ucsd.edu/speccoll/siooralhistories/Arrhenius.pdf>. Status: May 23, 2018.

Heil, Christiane: Mutter Natur entscheidet, wann wir löschen können *FAZ Online,* 6.12.2017. <http://www.faz.net/aktuell/gesellschaft/unglueck e/d-15327788.html>. Status: October 10, 2018.

Helmholtz-Gemeinschaft (ed.): Helmholtz-Roadmap für Forschungsinfrastrukturen II 2015, Korrigierte Version vom 21.9.2015, 17. <https://www.helmholtz.de/fileadmin/user_upload/publikationen/Helmholtz_Roadmap_2015_web_korr_150921.pdf>. Status: May 23, 2018.

Hofmann, Gustav: Mügge, Ratje. *Neue Deutsche Biographie* 18 (1997), 267 – 268 [Onlinefassung]. <http://www.deutsche-biographie.de/pnd133807681.html>. Status: May 23, 2018.

Homepage oft he Working-Group Earth System Sciences of the Leopoldina: <https://www.leopoldina.org/politikberatung/arbeitsgruppen/erdsystemforschung/>. Status: May 23, 2018.

Homepage of the Academy of Sciences and Arts, North Rhine-Westphalia, <http://www.awk.nrw.de/akademie/klassen/naturmedizin/ordentliche-mitglieder/ehhalt-dieter-hans.html>. Status: May 23, 2018.

Homepage of the German Museum in Munich: Willkommen im Anthropozän – Unsere Verantwortung für die Zukunft der Erde. <http://www.deutsches-museum.de/ausstellungen/sonderausstellungen/rueckblick/2015/anthropozaen/>. Status: May 23, 2018.

Homepage of the Global Fire Monitoring Center (GFMC), <http://gfmc.online/intro/About1.html>. Status: October 10, 2018.

Homepage of the Haus der Kulturen der Welt (HKW), <http://www.hkw.de/de/programm/projekte/2014/anthropozaen/anthropozaen_2013_2014.php>. Status: May 23, 2018.

Homepage of the Leibniz Institute for Tropospheric Research (TROPOS), <https://www.tropos.de/institut/ueber-uns/das-institut/>. Status: May 23, 2018.

Homepage of the MPI for Chemistry (MPIC), <https://www.mpic.de/ueber-uns.html>. Status: Februrary 7, 2018.
- Department for Atmospheric Chemistry: <https://www.mpic.de/forschung/atmosphaerenchemie.html>. Status: May 23, 2018.
- Department for Particle-Chemistry: <https://www.mpic.de/forschung/partikelchemie.html>. Status: May 23, 2018.
- Department for Multiphase-Chemistry: <https://www.mpic.de/forschung/multiphasenchemie.html>. Status: May 23, 2018.
- Department for Climate-Geochemistry: <https://www.mpic.de/forschung/klimageochemie.html>. Status: May 23, 2018.
- Research Overview of the MPIC, <https://www.mpic.de/forschung/uebersicht.html>. Status: May 23, 2018.

Homepage of the MPI for Meteorology (MPI-M), <https://www.mpimet.mpg.de/en/institute/organisation/>. Status: May 23, 2018.

Homepage of the Museum of Technology Sinsheim and Speyer <http://www.bredow-web.de/Sinsheim-Speyer/Tupolev_TU_144/tupolev_tu_144.html>. Status: May 23, 2018.

Homepage of the " Staubquellen", <http://www.tropos.de/institut/abteilungen/modellierung-atmosphaerischer-prozesse/transportprozesse/staubquellen/>. Status: May 23, 2018.

Homepage of Josephine Wall, Gaia-paintings <http://josephinewall.co.uk/art-gallery/goddesses/>. Status: September 19, 2018.

Jaenicke, Ruprecht: Laudatio auf Kurt Bullrich, anlässlich seines Todes am 31.3.2010. <https://www.blogs.uni-mainz.de/fb08-ipa/files/2014/07/Laudatio_bullrich.pdf>. Status: May 23, 2018.

Klaesgen, Michael: Wilde Jagd am Himmel. In Paris zerschellte der Mythos der Sicherheit. *Zeit Online*, 27.7.2000. <https://www.zeit.de/2000/31/Wilde_Jagd_am_Himmel>. Status: May 23, 2018.

Libby, Willard F.: "Radiocarbon dating." *Nobel Lecture* 12.12.1960. <https://www.nobelprize.org/uploads/2018/06/libby-lecture.pdf>, status: October 9, 2018.

Homepage oft he Anthropocene-Working Group, <http://quaternary.stratigraphy.org/working-groups/anthropocene/>. Status: October 10, 2018.

Mößbauer, Rudolf: Rudolf Mössbauer—Biographical. <https://www.nobelprize.org/prizes/physics/1961/mossbauer/biographical/>. Status: October 9, 2018.

Nobel Prize Organization, Press Release „The Nobel Prize in Chemistry 1995". <https://www.nobelprize.org/prizes/chemistry/1995/press-release/>. Status: May 23, 2018.

Prelowski Liebowitz, Ruth: Landsberg, Helmut Erich. <http://www.encyclopedia.com/doc/1G2-2830905841.html>. Status: May 23, 2018.

Priesner, Claus: Schönbein, Christian Friedrich. <http://www.deutsche-biographie.de/sfz78953.html>. Status: May 23, 2018.

Rudstam, Sven Gösta: Who's who in CERN. <http://lib-docs.web.cern.ch/lib-docs/Archives/biographies/Rudstam_G-196303.pdf>. Status: May 23, 2018.

Schaaf, Michael: Schweres Wasser und Zentrifugen. Paul Harteck in Hamburg (1934 – 1951), URL: <http://censis.informatik.uni-hamburg.de/publications/Art_M_Schaaf_Harteck.pdf>, October 1, 2014. Status: October 9, 2018.

Schmidt-Rohr, Ulrich: Wolfgang Gentner. 1906 – 1980". <https://www.leo-bw.de/web/guest/detail/-/Detail/details/PERSON/kgl_biographien/118538470/Gentner+Wolfgang>. Status: May 23, 2018.

Schwägerl, Christian: Planet der Menschen. *Zeit Online*, 18.2.2014. <http://www.zeit.de/zeit-wissen/2014/02/anthropozaen-planet-der-menschen>. Status: May 23, 2018.

UN, Online-Archiv: General Assembly. <http://www.un.org/en/ga/search/view_doc.asp?symbol=A/43/351>. Status: May 23, 2018.

Urey, Harold: Some thermodynamic properties of hydrogen and deuterium. Nobel Lecture 14.02.1935. <https://www.nobelprize.org/uploads/2018/06/urey-lecture.pdf>. Status: October 9, 2018.

Weart, Spencer: The Discovery of Global Warming. <https://history.aip.org/climate/>. Status: October 9, 2018.

Abbreviations

ABLE-2A	Amazon Boundary Layer Experiments
AFCRC	Air-Force-Cambridge Center
AMPG	Archiv der Max-Planck-Gesellschaft
CalTech	California Institute of Technology
CERN	Conseil Européen pour la la Recherche Nucléare
CIAP	US-Climate Impact Assessment Program
CITE-3	Chemical Instrumentation Test and Evaluation-Expedition
CLAIRE 98 und 2001	Cooperative LBA Airborne Regional Experiment
CPT-Sektion	Chemisch-Physikalisch-Technische Sektion der MPG
DFG	Deutsche Forschungsgemeinschaft
DMSP	US-Defense Meteorological Satellite Program
ENUWAR	Environmental consequences of Nuclear War
ESS	Earth-System-Sciences
ESSP	Earth-System-Science Partnership
EURATOM	Europäische Atomgemeinschaft
FCKW	Flourchlorkohlenwasserstoffe (Sammelbegriff)
FOS/DECAFE-91	Fire of Savanna/Dynamique et Chimie Atmosphèrique en Forêt Equatoriale
FRG	Federal Republic of Germany
GDCh	Gesellschaft Deutscher Chemiker
GFMC	Global Fire Monitoring Center
GWFN	Global Wildland Fire Network
HGF	Helmholtz-Gemeinschaft
ICS	International Commission on Stratigraphy
ICSU	International Council of Scientific Unions
IGAC	International Global Atmospheric Chemistry Programme
IGBP	International Geosphere-Biosphere Program

IKO	Instituut voor Kernfysisch Onderzoek Amsterdam
IMG	Frankfurter Institut für Meteorologie und Geophysik
INF	Intermediate Range Nuclear Forces
IPCC	Intergovernmental Panel on Climate Change
IUGS	International Union of Geological Sciences
KFA	Kernforschungsanstalt Jülich
KWIC	Kaiser-Wilhelm-Institut für Chemie
MISU	Meteorologisches Institut der Universität Stockholm
MPG	Max-Planck-Gesellschaft
MPI	Max-Planck-Institut
MPI-BGC	Max-Planck-Institut für Biogeochemie
MPIC	Max-Planck-Institut für Chemie
MPI-M	Max-Planck-Institut für Meteorologie
NASA	US-National Aeronautics and Space Administration
NCAR	National Center for Atmospheric Research
NSDAP	Nationalsozialistische Deutsche Arbeiterpartei
PSC	Polar Stratospheric Clouds
SA	Sturmabteilung
SAFARI 92	Southern African Fire-Atmosphere Research Initiative
SCOPE	Scientific Committee on problems of the environment
SDI	Strategic Defence Initiative
SFB	Sonderforschungsbereich
SST	Super Sonic Transport
TROPOS	Institut für Troposphärenforschung
UNISDR	UN International Strategy for Disaster Reduction
WCRP	World Climate Research Program
WMO	World Meteorological Organization

INDEX

Italic page numbers refer to footnotes, **bold** to legends.

F

G

H

J

K

L